The Decline of the Daily Newspaper

Digital Formations

Steve Jones
General Editor

Vol. 83

The Digital Formations series is part of the Peter Lang Media and Communication list.
Every volume is peer reviewed and meets
the highest quality standards for content and production.

PETER LANG
New York • Washington, D.C./Baltimore • Bern
Frankfurt • Berlin • Brussels • Vienna • Oxford

KEITH L. HERNDON

The Decline of the Daily Newspaper

How an American Institution Lost the Online Revolution

PETER LANG
New York • Washington, D.C./Baltimore • Bern
Frankfurt • Berlin • Brussels • Vienna • Oxford

Library of Congress Cataloging-in-Publication Data

Herndon, Keith L.
The decline of the daily newspaper: how an American institution
lost the online revolution / Keith L. Herndon.
p. cm. — (Digital formations; v. 83)
Includes bibliographical references and index.
1. Newspaper publishing—United States—History—21st century.
2. Newspaper publishing—Technological innovations—United States.
3. Electronic newspapers—United States—History—21st century. I. Title.
PN4867.2.H47   071'.309051—dc23   2012010517
ISBN 978-1-4331-1720-6 (hardcover)
ISBN 978-1-4331-1974-3 (paperback)
ISBN: 978-1-4539-0870-9 (e-book)
ISSN 1526-3169

Bibliographic information published by **Die Deutsche Nationalbibliothek**.
Die Deutsche Nationalbibliothek lists this publication in the "Deutsche
Nationalbibliografie"; detailed bibliographic data is available
on the Internet at http://dnb.d-nb.de/.

Cover design and artwork by Marie Matthews,
an artist and technologist who lives in Atlanta, Georgia.

The paper in this book meets the guidelines for permanence and durability
of the Committee on Production Guidelines for Book Longevity
of the Council of Library Resources.

This book is dedicated to Dr. Matthew Allen,
a trusted adviser and true friend.

Contents

Introduction   1

1   Videotext and the Birth of Online Newspapers   23
2   The Newspaper Industry's Brief Cable Television Strategy   61
3   Newspapers React to Fear of Telecommunication Dominance   85
4   Newspapers Embrace Proprietary Online Services   109
5   The Emerging Internet Threatens the Established
    Publishing Model   133
6   Mergers, Convergence, and an Industry under Siege   173
7   Connecting the Lessons of History   207

    Conclusion   227
    Notes   231
    Bibliography   267
    Index   291

Introduction

A Pittsburgh newspaper published a column in 1903 contemplating what newspapers would be like in the year 2000. The author's forecast could not conceive of the technology newspapers would confront considering that radio and television were yet to be introduced, but he speculated that news would "be followed and reported, irrespective of distance."[1] The columnist also correctly surmised that readers in the 21st century would have little interest "in what the editor thinks."[2] But the most relevant aspect of this column was not something written, but implied. The turn-of-the-century futurist never questioned the newspaper industry's survival. He simply took for granted that newspapers would be around in that distant future.

This proved to be a valid assumption. The newspaper industry has survived the more than 100 years since that column appeared. Should such a column be written today, however, the author would face the future of the newspaper industry with much less confidence. The success of the Internet and the newspaper industry's own missteps have caused industry executives and analysts, journalists, and scholars to re-examine the position of the printed newspaper within the media landscape. Newspaper companies have struggled to remain competitive in the online era and faced deteriorating financial trends—precarious circumstances made worse by a deep economic recession in the late 2000s. Newspaper companies—having lost nearly half of their advertising revenue between 2005 and

2010[3]—now search for new business models in the online and mobile arenas in an effort to stave off extinction.

This book examines and reports from a historical perspective the issues and forces that shaped the U.S. newspaper industry during the online era, a period that began in 1980 and continues through to the present. Collectively, these three-plus decades represent a period of extreme change and transition. This book seeks to explain how the decisions made during the early part of this period profoundly influenced the newspaper industry's current circumstances.

Looking back from today, the online era can be understood as a period that began when the Information Age transformed from rhetoric into tangible products and services. Digital computer technologies emerged and were married with telecommunications infrastructure to form information networks. The early closed, proprietary systems demonstrated the potential of computer-based communications. However, the open architecture of the Internet represented a paradigm shift that altered the balance of power between information providers and their consumers. This book explores the years leading up to the time when newspaper publishers first confronted the Internet, the years when the Internet was adopted by newspapers as a distribution platform, and the years immediately following that period when an Internet-based media economy emerged to the detriment of newspaper business models.

Throughout this period, there has been an ongoing analysis of the newspaper industry's past and speculation about its future. Material gleaned from industry magazines, the mainstream press, and scholarly journals provides a constant commentary that reveals an industry anxious about what always seemed to be described as a precarious future. The industry's self-doubts, however, were tempered in the 1980s and 1990s by the successes of a business model that generated large profits even as underlying business fundamentals such as the share of the advertising market and circulation were declining. The variety of material available demonstrated that the newspaper industry in the United States of America has a tradition of introspection. From an academic research perspective, this tradition is manifested within the disciplines of media history and media industry studies, both of which have been influenced by the work of Marshall McLuhan.

McLuhan's Influence

Sometimes described as the father of media studies, McLuhan is an essential figure in the discussion of media history and represents key underlying constructs of this book. He is recalled for inextricably linking human history with the development of communication technologies over time. Sparks explained that McLuhan "claimed that the invention of print brought about cataclysmic change in human culture and that the invention of electronic media started a major revolution that we have yet to complete."[4] McLuhan did not live to see the emergence of the Internet, but his observation that "the medium is the message"[5] is viewed as a "prophetic vision" of how the Internet blurred the lines between content creation and its distribution.[6]

McLuhan, as Sparks noted, "never did content analyses, surveys or experiments to test his ideas."[7] His contribution to media history as a discipline is derived from the concept that great insight can be achieved by studying qualitatively what has come before. Hodge observed that McLuhan's "initial object of study was not the revolutionary coming of the new electronic media but the previous revolution, the coming of print and its 400 years of dominance."[8] Hodge noted that "McLuhan's grand narrative of the history of media took place against the background of a conventional history," which provided a framework for his theories.[9] McLuhan's specific influence on this book is found in Brooke's observation that "McLuhan routinely criticized our culture for forcing new media into doing the work of the old."[10] The inherent tension that arises from such activity is explored throughout this volume as the newspaper industry tried to forge a relationship with online media. McLuhan stated: "When faced with a totally new situation, we tend always to attach ourselves to the objects, to the flavor of the most recent past. We look at the present through a rear-view mirror. We march backwards into the future."[11] For media history, McLuhan's words are a guide for not only interpreting the topic of study, but in producing the history itself. His work illustrates how to discern and comment on future media developments by rooting those observations in what can be gleaned from previous media developments.

Historical Perspective

In this light, media history research can be deployed as a tool for use in broader media and new media studies. Peters commented that "the strength of new media studies and media history lies in their merger."[12] Media history is not without the same shortcomings that are inherent in any history. Kyrish observed that "a major difficulty in learning from the past is that we already know how it turned out."[13] She cautioned that historical reviews can be limited because "implications of past actions can only be confirmed through—and because of—the passage of time," explaining further "that knowing about past outcomes makes it extremely difficult not to see those outcomes today as inevitable and therefore knowable."[14] The media historian must strive to see the period reviewed through the lens of those living the events and understand that those participants did not have the benefit of complete foresight.

Regarding this point, Stöber discussed the limitations of those living in the moment of what is later studied as history. He wrote: "Suppose contemporaries of Gutenberg had been asked about the social consequences of Gutenberg's invention" of moveable type. He concluded that "no contemporary of Gutenberg would have had any chance to imagine newspapers and magazines."[15] This same concept applies to modern inventions. People at the start of the personal-computer era did not grasp the extent to which the device would affect society. Kaletsky noted that "even as late as 1980, no one would have put any significant probability on computer sales exceeding car sales by a factor of ten to one."[16] These examples—the invention of moveable type and the introduction of the personal computer—are two of the most significant events in media history. Understanding that the full comprehension of those innovations eluded their contemporaries underscores the burden of media historians. When examining human reaction to events and circumstances, it becomes imperative to account for context.

Researchers cannot escape the fact that current knowledge serves to bias and prejudice history. Researchers must accept the reality that as knowledge changes over time, so does the understanding and interpretation of history. Peters noted this, stating that "the way in which scholars understand history colors and shapes the evolving and contingent enterprise of understanding media."[17] Even with that caveat, however, Peters believed that media history serves as an important component in understanding the development of new media, which makes it an integral part of new media studies as a discipline. He wrote: "to study new media is to ride squarely atop the ever-unfolding crest between the past and the present."[18] This book, however, is not new-media history per se; it was approached from the perspective of the newspaper industry. Therefore, it is a narrative history of the newspaper industry's reaction to the emergence of new media and chronicles the newspaper industry during a period of dramatic transition. By approaching the newspaper industry as media confronting transition, the book relies on the notion expressed by Gitelman that "looking into the novelty years, transitional states, and identity crises of different media stands to tell us much, both about the course of media history and about the broad conditions by which media and communications are and have been shaped."[19]

Media Industry Perspective

In considering the newspaper business collectively as an industry, this book identifies with emerging scholarship specific to media industry studies. Media industry studies can be understood as a subset of the broader cultural industries genre and considers history as a significant component of the discipline.[20] As a media history undertaken in the spirit of media industry scholarship, this book was inspired by the works of Winseck and Light. Winseck examined the early telegraphy era between 1840 and 1910 in Canada, Britain, and the United States, while Light explored facsimile as a 20th-century medium. Winseck wrote that "contemporary discussions of new media, information services and convergence proceed if these are entirely new phenomena" but noted that his research found "a similar pattern of events" in the history of the telegraph industry.[21] Similarly, contemporary discussion of the newspaper industry's relationship with online media often begins with the emergence of the Internet, but this book examines a much longer and more complex history.

Winseck's work revealed that "newspaper publishers were among the first and most generous investors in setting up telegraph operators, providing the largest source of revenue, and establishing organizations . . . to exploit the potentials of electronic communications."[22] Just as Winseck recounted the newspaper industry's efforts in telegraphy, this book does the same for an era that began 70 years later. And it will attempt to deliver on the same goal: "If recognizing historical patterns provides any guidance for the present, perhaps the history of electronic communication presented in this paper can provide some insights into the nature of media evolution today."[23] Such work creates a forum for exploring how media industries react to forces of change and for understanding how those reactions are manifested in product decisions over time.

Light wrote that the field of "new media studies, with its frequent comparative historical focus, opens a door for scholars to revisit and to question the historiographical boundaries of the old media studies—what technological systems have received disproportionate attention, and what new histories of old media might be written."[24] Although this book is written from the perspective of an old medium, it honors Light's premise by explaining how technologies such as videotext and cable television were deployed by newspaper companies in a quest to exploit online media. As Light found with the gaps in media history regarding facsimile, the discussion of the newspaper industry's online experience is often limited to the Internet era—downplaying the fact that the newspaper industry's relationship with online media had begun more than a decade earlier.

Gitelman suggested that media history address broad questions that "have practical ramifications."[25] This list of questions included:

> Is the history of media first and foremost the history of technological methods and devices? Or is the history of media better understood as the story of modern ideas of communication? Or is it about modes and habits of perception? Or about political choices and structures? Should we be looking for a sequence of separate "ages" with ruptures, revolutions, or paradigm shifts in between, or should we be seeing more of an evolution? A progress?[26]

When considering these questions as they applied to this book, a framework emerged that examines newspaper companies through the choices they made and the structures they created in response to significant technology developments. The actions of one newspaper company can be separate and distinct from the actions of another newspaper company, but they are nevertheless members of the same industry. Therefore, in keeping with the media industry construct, this book examines the commonality among newspaper responses and treats these collectively as the newspaper industry response.

Holt and Perren recognized that media industry study is interdisciplinary, drawing from work undertaken in many fields including "film and television studies, communication, law, public policy, business, economics, journalism and sociology." They also promoted the notion that relevant material can be gleaned "far beyond the traditional purview of academic study," noting that "discourses in the trade papers, the popular press, and academic publications are supplemented by writing in digital communities, online journals and the blogosphere."[27] This book follows such an interdisciplinary approach and has drawn material from all of these sources. Holt and Perren asserted "this range of perspective is both a necessary component and a constitutive element of" media industry study.[28]

The efforts of dozens of scholars also are featured in this book. The works of Pablo J. Boczkowski and Sandy Kyrish, for example, were essential to defining its historical framework. Boczkowski's influence deserves a special mention given that he observed that a "misunderstanding . . . quite pervasive in both academic and popular discourse" is that the newspaper industry's move to the Internet "was some sort of revolutionary occurrence and without any roots in the past." He argued that "it was a far more evolutionary process influenced by a history of tinkering with multiple forms and many facets of consumer-oriented electronic publishing."[29] This book corroborates this line of reasoning as it illustrates how deliberate the newspaper industry's approach to online delivery actually was.

Within a media industry perspective, it is necessary to be clear about what is meant by the phrase "newspaper industry." Smith explained that "one writes 'the newspaper,' but of course the newspaper exists in three or four quite different varieties."[30] Smith was counting nationally circulated newspapers such as the *New York Times* and the *Wall Street Journal* as well as metropolitan and small-town dailies. There are other types of newspapers such as weeklies and those focused on specific topics such as a community within a larger city, entertainment, or regional business news. For the purposes of this book, however, the term "newspaper industry" is used to refer to the collection of companies that publish daily newspapers either nationally or regionally.

To create a foundation for the examination and analysis contained in this book, various statistics and rhetoric are presented that explain the condition of the newspaper industry as it is defined here.

Industry Conditions

The newspaper industry at the beginning of the 1980s was very prosperous, with overall revenue posting several years of double-digit growth. Total advertising revenue had climbed from $5.7 billion in 1970 to $14.8 billion in 1980.[31] The number of newspapers had remained essentially flat during the 1970s, and total circulation also remained flat at around 62.2 million daily copies. Sunday circulation, however, had increased from 49.2 million copies in 1970 to 54.7 million in 1980.[32] The newspaper industry added 49,000 jobs in the latter half of the 1970s, bringing the total in 1980 to 432,000 workers.[33] The record employment levels provided a strong boasting point for the industry's lead trade group: "With this latest spurt, newspapers passed steel mills, automobile/car body plants and auto parts manufacturers to assume the lead among the biggest U.S. manufacturing employers."[34]

It was a remarkable post-war performance. Radio had emerged as a national media phenomenon in the late 1920s creating new competition, and the Great Depression ravaged the newspaper industry in the 1930s. Between 1929 and 1933, for example, newspaper advertising revenue declined 45 percent. Several hundred newspapers ceased publication, which forced thousands of employees out of work.[35] The newspaper industry recovered from the Depression era when it was buoyed by the demand for information during World War II. It had withstood the introduction of commercial radio, the rapid expansion of television in the 1950s and 1960s, and the stagnant economy of the 1970s. It entered the 1980s acting as the formidable media industry competitor it was.

In January 1980, Allen H. Neuharth, chairman and president of Gannett Inc. and the leader of the industry's trade association, described the coming decade as one of "challenge and opportunity for newspaper people." He summarized three distinct trends:[36]

> There will be more diversity of news and views than ever, and more competition too. There will be more leisure time than ever for our audiences, and the interests of those audiences will be more specialized. There will be more new technology, which will give us the means to better fill the needs of those audiences.[37]

This book is particularly concerned with Neuharth's third trend, in which he anticipated an increased use of technology. The statement demonstrated the industry's early belief that its mission included deploying new technologies— whatever they may be—in order to serve its readers. Neuharth's comments illustrate how newspaper industry leaders were adopting the rhetoric of the day and making it their own. The basic premise of a technology-driven societal revolution became a common theme in popular literature and culture during the early 1980s. The revolutionary rhetoric reflected in the works of such authors as Alvin Toffler[38] and John Naisbitt[39] espoused a coming period of significant technological change. The industry's early rhetoric acknowledged that new technology represented potential new competition, but it was expressed in ways that reflected confidence that newspaper companies would accept the challenges and adapt to new market conditions. Consider the following excerpt from an essay published in early 1980 by the industry's trade association:

> The new technology . . . will change our business more drastically than the conversion from hot-metal printing or any other technological evolution that we've experienced. But unlike those evolutions, if we don't take advantage of these new opportunities, those outside the newspaper business will. Many people believe that the 1980s will deliver society into the "The Information Age." If newspapers are alert, new technology has taken us to the threshold of a business limited only by our imaginative and creative capacities.[40]

From this point, newspaper companies engaged in numerous activities as they attempted to exploit new technology or defend against it.

The book follows the newspaper industry throughout this period as its companies launched and cancelled videotext projects, entered and abandoned cable television partnerships, lobbied to keep the telecommunications industry out of online information, partnered (briefly) with proprietary online services, and, finally, confronted the emergence of the Internet. Examining these topics as distinct milestones in the history of the newspaper industry provides insight and understanding about the industry as its underlying business model deteriorated throughout the period.

The industry's early efforts with videotext were deemed failures by many when they proved expensive relative to the audience they attracted. In search of a new online strategy, newspaper companies turned to partnerships with proprietary online systems that emerged in the late 1980s and early 1990s. But these efforts, too, were criticized as limited in scope. Critics contended that newspaper companies needed to do more to secure their future. Maney summarized this perspective: "The newspaper industry is a deer frozen in megamedia's headlights, soon to be creamed if it doesn't get moving."[41] Rather than bold initiatives, however, newspaper companies committed to their online alliances with companies such as Prodigy and America Online (AOL) until the Internet erupted.

When newspaper companies migrated to the Internet as an online platform, they confronted many of the same issues that had bedeviled earlier online projects. This book examines the newspaper industry's approach to the open structure of the Internet and explores how the Internet led to new investments, partnerships, and acquisitions and caused the industry to reconsider fundamental business processes and procedures.

Advertising revenue for printed newspapers peaked in 2000, and the following years found the industry struggling with recession and increasing online competition.[42] The industry's advertising growth had lagged behind other media during the 1990s. In the mid-1990s, television—led by a surge in cable advertising—supplanted newspapers as the nation's largest advertising medium.[43] The hope that online revenue would offset declines in printed revenue failed to materialize. By 2005, newspaper companies generated slightly more than $2 billion in online revenue, which allowed the industry to achieve a new overall revenue peak of $49.4 billion. Over the next five years, however, newspaper companies grew their online revenue by only $1 billion, reaching slightly more than $3 billion in 2010. However, during this same period, revenue from advertising in printed newspapers collapsed, falling $24.6 billion.[44]

The erosion in advertising market share tracked declining circulation that had plagued newspaper companies for decades. As they had in the 1970s, newspapers managed to keep circulation relatively stable during the 1980s. But by 2005, total daily newspaper circulation in the U.S. was 53.3 million copies, a decline of more than 14 percent in 15 years.[45] The circulation erosion continued to around 45 million copies by the end of the decade, representing another 15 percent decline in five years.[46] From 1980 to 2008, 337 daily newspapers ceased publication.[47]

As revenue declined and the number of newspapers dwindled, so did the industry's total employment. The industry had remained a job creator throughout the 1980s, with employment peaking at nearly 486,000 jobs in 1991. In subsequent years, however, the industry lost thousands of jobs due to production automation, expense controls, and the decreased number of newspapers. Employment was around 372,000 workers in 2006[48] and declined to 326,000 in 2008.[49] That year, the U.S. Bureau of Labor Statistics projected that newspaper employment would fall to 245,000 workers by 2018.[50] At the end of the decade, the forecasted decline was on track as industry jobs shrank to about 259,000 by early 2010.[51]

When Knight-Ridder announced at the end of 2005 that it was selling its newspapers and closing the company, industry executives understood that their business had fundamentally changed. Newspaper companies had managed some success in attracting online audience, but they had no clear strategy for monetizing that audience, and their business remained dependent on the printed product even though it was losing ground in the marketplace. In looking back to this period, 2005 represented a watershed moment. It was the point in the industry's history where tangible evidence—not only pundits' prognostications—suggested that the newspaper industry's business cycle had run its course.

Predictions of Demise

There had long been predictions that computer technology would lead to the industry's demise. Ted Turner, the founder of Turner Broadcasting and its Cable News Network (CNN), forecasted in 1980 that the daily newspaper industry would not survive the decade. In 1990, Turner admitted that his timing was off, but reiterated his basic prophecy: "It may take 10 or 20 more [years]. Newspapers will eventually go."[52] By the mid-1990s, prominent industry insiders also were speculating about the end of printed newspapers. A Knight-Ridder executive working with new electronic platforms said print would be replaced with electronic editions by 2005, and a former editor turned Internet consultant predicted newspapers "will disappear over the next 15 to 20 years."[53]

Although the timing of these predictions did not come true, the increase in negative discourse about the future of printed newspapers underscored how the industry's reputation had shifted from Information Age pioneer in the early 1980s to flailing follower by the early 1990s—even before the Internet had emerged as a powerful new force in the media marketplace. Popular author Michael Crichton described newspaper companies as belonging to the "mediasaurus," the soon-to-be extinct mass media, and charged that they were not responding effectively enough to the challenges of new technologies to survive.[54] Tom Peters, a leading management consultant, speaking at a newspaper marketing conference said "if this is the age of information, [then] this should be the dawning of a great era for newspapers, not their eclipse." But he concluded with an observation that the newspaper industry had a history of acting too cautiously, stating "I'm not sure that it's up to that level of craziness that's required for survival."[55] After decades of success and media industry leadership, newspaper executives were unaccustomed to such public criticism.

In a lengthy rebuttal published in the newspaper association's trade journal, an editor wrote of traditional media: "Newspapers and networks may have been slow to react to the Information Age," but "they will ultimately be its architects."[56] This comment is illustrative of the newspaper industry's collective belief throughout much of the online era that it would figure out how to succeed eventually. The attitude reflected the sensibilities of a profitable mature industry, but it also led to internal tension regarding the strategies and tactics that were undertaken in pursuit of online success. As the industry attempted to exploit opportunities in the online era, it struggled to define the inherent nature of the newspaper industry's business.

This manifested itself in two ways. First, newspaper executives were challenged with thinking about their business holistically as information, not printed newspapers. Second, as negative advertising and circulation trends began to affect financial performance, newspaper companies engaged in a series of expense cuts and staff layoffs. This erupted into open discord between the industry's business management and many of its editors and journalists.

The Nature of the Business

As early as 1980, newspaper industry critics were cajoling executives to think about a broader market—an information marketplace rather than the specific product called a newspaper. It was similar to the horse-and-buggy analogy used when the car industry began. Were horse-drawn-carriage makers only in the carriage business or were they part of a larger transportation industry? Would newspapers suffer the same fate as the carriage makers who refused to adapt to the encroachment of Henry Ford and the Model T? In 1980, Compaine framed the industry's choices in defining its business as a series of questions:

The most important answer here must come to the question, "What is a newspaper?"

Is it a format called "ink on newsprint"?
Is it a delivery method of private carriers and newsstands?
Is it a package of information?[57]

Compaine answered his questions by asserting that newspapers were "essentially an information package, one that just happens to be printed on newsprint for now." He argued that a newspaper company must be viewed as "a collector and disseminator of information" and suggested the industry "stop using the phrase 'newspaper' and think of the business in information terms," concluding that "it would be myopic to lose sight of the less-obvious competitors, which today may be known as banks, computer companies and catalog mail-order firms—or other creatures for which there is as yet no name."[58] Compaine's observations turned out to be very prescient as the unnamed creatures he worried about in 1980 became formidable media competitors with names such as Amazon, eBay, Yahoo, and Google.

In the beginning of the online era, however, the newspaper industry was not all that concerned with its identity. There was a widely held belief among executives that the industry could draw from its historical experience with earlier electronic media—television and radio—as inspiration and guidance. Newspaper companies had been deeply involved in the early development of television and radio, especially in providing news content for these media. A trade journal wrote that the newspaper industry's contributions to earlier electronic media "belie the parochial view that newspapers are modern-day luddites bent on impeding, if not destroying, any technology that threatens their ink-on-paper products."[59] A newspaper executive recalled the early development of television when newspaper newsrooms had television cameras on reporters' desks and easels for displaying newspaper pages on camera. "Nobody knew what television was in 1947," he said, asserting that the online era warranted the same types of experiments. He added that newspaper companies had to learn if online media "is so different that it develops into entirely independent organizations [as broadcast did], or whether there's enough similarity between newspaper editing that our organizations can evolve internally into hybrids."[60] An important aspect of this book deals with how newspaper companies addressed these concerns through organizational structures and other business considerations.

A discussion of newsgathering and the presentation of news as it relates to how newspaper companies dealt with online content is also an important element of this book. The advent of interactive technologies challenged many of the newspaper industry's traditions, processes, and procedures. As a book focused on the *business* of newspaper companies as commercial enterprises, it does not analyze their actions from a journalistic perspective. However, understanding the newspaper industry requires an appreciation for its journalistic sensibilities.

A Noble Institution

Bovee said "the function of journalism is to provide people, individually and as members of communities, with the knowledge that will help them make good, timely decisions about what should or should not be done."[61] No part of that definition ties journalism to any particular product form. Newspaper people, however, linked their product very closely to the definitions of journalism.

The significance of the journalistic mission is fundamental in how the industry thinks about itself and how it responds to challenges. Udell maintained that newspaper companies occupy a unique position in American industry: they are dependent on the free enterprise system for their livelihood, but they are protected by the free press provisions of the U.S. Constitution. He wrote:

> It was in this context that the American newspaper developed; protected first in its right to seek truth regardless of the path down which truth leads; and second—as free private enterprise—motivated to profit by satisfying the needs of its customers.[62]

Journalists want their newspaper employers to operate as noble institutions at the vanguard of a constitutionally guaranteed free press. Publishers are fine with that perspective as long as profit margins are high enough to ensure economic freedom is not compromised. From the perspective of investors, however, the discussion about journalistic mission and product quality was much ado about nothing. Warren Buffett, a prominent investor, asserted in the early 1980s that product quality may have contributed to a newspaper's local market dominance; but once dominant in a market, quality no longer was a factor in ongoing profitability. He wrote:

> The economics of a dominant newspaper are excellent, among the very best in the business world. Owners, naturally, would like to believe that their wonderful profitability is achieved only because they unfailingly turn out a wonderful product. That comfortable theory wilts before an uncomfortable fact. While first-class newspapers make excellent profits, the profits of third-rate papers are as good or better. . . .[63]

The disassociation of newspapers as unique societal institutions and newspaper companies as profit-making businesses became more pronounced in the online era.

As negative industry trends began to accelerate, publishers implemented stringent expense controls to maintain profit margins. The ensuing layoffs and budget cuts did not spare the newsrooms, which led to open dissent from journalists. Editors and reporters questioned the motives of their business managers, expressing deeply seated beliefs the industry was compromising journalistic integrity in favor of profits. A leading industry analyst wrote that the business was "under siege,"[64] while the ombudsman at the *Washington Post* described the industry in the mid-1990s this way:

> We have publishers under enormous pressure to produce profits who are afraid to tell their newsroom people what the situation is. We have editors buffering reporters from the realities even as they bring about the cuts needed to protect profits. And we have journalists unwilling to hear these business truths and famously allergic to change.[65]

These comments underscored how difficult it was for newspaper companies—the business managers as well as the journalists—to adapt to a new marketplace. Throughout the period discussed in this book, comments and examples of activities illustrated that newspaper companies were complex organizations steeped in tradition and conservative business practices.

Through its expository approach, this book allows for the story of the newspaper industry during this period to unfold as the trajectory of online media increases. Given the long history of newspapers as a media form, this book examines events that occurred during a relatively short period of time. However, the significance of the changes that occurred during this period cannot be understated. For example, Buffett, the billionaire investor who was very positive about the industry in the early 1980s, had an entirely different outlook by the late 2000s. He stated: "For most newspapers in the United States, we would not buy them at any price. They have the possibility of going to just unending losses." The difference was that newspapers "were once essential to the American public" but lost that prominence to a variety of competing sources for news content including the Internet.[66]

Nevertheless, Buffett in late 2011 spent $200 million to acquire the *Omaha World-Herald*, the daily newspaper in his hometown. Buffett said it was a newspaper that "delivers solid profits," but an analyst said the motive behind the purchase was not financial, but altruistic: "It's his hometown paper so he probably wants to see it preserved."[67] A billionaire buying a newspaper out of nostalgia underscores the industry's plight. Such a dramatic fall in the industry's standing explains the importance of studying the newspaper industry during the digital

era. A better understanding of how new technologies challenged newspaper owners and managers and contributed to the collapse of a long-established business
model is necessary if the industry hopes to learn from its mistakes and find pros-
perity in today's new media landscape dominated by mobile communications.

Key Concepts and Definitions

A book of this nature cannot be presented to its reader without attempting to
explain the conceptual basis on which interactive technology was developed and
understood—and in some cases misunderstood—in the newspaper business during the online era. The phrases "new media" and "interactive media" have usage
implications, as does the term "convergence." Therefore, understanding the conceptual framework of these terms will help the reader appreciate how they are used
within the context of this book.

New Media

The term "new media" emerged as a way to illustrate a difference in the media
it described in contrast to "traditional media," which generally include books,
newspapers, magazines, and television, as well as recorded music and radio. Mayer
wrote that "new media" should be considered as a phrase relative to the time of
its usage. "At one time, the printing press would have been considered a new
medium," he wrote, adding that radio and television also were considered new
media when they debuted.[68] Mayer stated that the phrase "in its most recent
incarnation" had been "adopted to refer to a series of scientific and technological
innovations . . . that led to the development of a considerable number of new
methods for creating, transmitting, and storing information and to the large-scale
transformation of many of the more traditional media."[69]

Rice[70] and Manovich[71] associated the phrase with media that exists in digital form. For a time in the mid-2000s, the phrase "new media" seemed to have
narrowed further to mean media specifically delivered through a connection to
the Internet. As such, Koziol observed that "new media is not so new anymore,"
considering that the Internet as a consumer platform is more than a decade old.
The phrase retains its currency, he wrote, because "perhaps it is 'young' relative
to more traditional media such as television, print and radio. . . ."[72] The concept
of new media has changed again, however, with the proliferation of wireless platforms that provide consumers with new ways to engage with media content. In

their book *Mobile Communications: An Introduction to New Media*, Green and Haddon, for example, directly address mobile media as a form of "new media" that is having a profound effect on global cultures and societies.[73] Therefore, in the context of mobile media, the term "new media" has found new life.

Interactive Media

Within this book, "interactive media" is often used as an attempt to more accurately describe the online competition the newspaper industry faced. In the online era, digital media was not only "new," but also "interactive" in that it provided an audience with a connection to the content creators and to other members of the audience as well. Nevertheless, scholars have found that defining "interactive media" is a complex task. The phrase is fraught with ambiguity because the underlying concept of interactivity is broadly applied, and often in incongruous ways. Before tackling how to explain "interactive media," it is necessary to briefly review the concept of "interactivity" and how it has been applied to the study of media.

Kiousis explained that "academic usage of 'interactivity' is marginally inconsistent at best."[74] He maintained the inconsistent usage was due to the many academic perspectives from which the concept has been studied, including media and communication, psychology, sociology, computer science, and information design.[75] Downes and McMillan noted that "scholars have employed the term [interactivity] to refer to everything from face-to-face exchanges to computer-mediated communication."[76] They determined that "much of the literature, both popular and scholarly, uses the term 'interactivity' with few or no attempts to define it."[77] Downes and McMillan attempted their own definition, but found that there was no single way of describing it, concluding that "varying levels of interactivity exist" and that application of the term depends on such factors as user perception, the timing and direction of communication, the responsiveness of the communication and the levels of control a user is allowed to exert over the process.[78]

This book embraces the perspective of "'interactivity' simply as a communication process,"[79] one that is "consistent with a human-computer interaction (HCI) approach" and "constructs 'interactivity' as a product of a medium characteristic…."[80] This is a straightforward approach that makes it clear the type of interactivity under discussion when specifically addressing the newspaper industry and how it attempted to exploit interactive technologies. Bucy's work underscores the acceptance of this approach in terms of defining interactivity:

Limiting the concept to exchanges that are in some way mediated by technology begins to distinguish the term from *any* form of communication and discourages its wanton application as a universal descriptor of all forms of dialogue.[81]

In this context, technology becomes the means of distribution for the media. In other words, the technology links the end user to the media and by that very act influences the user's experience as an interactive one. Traditional media, presented in its original form, must use external technology, such as email or the telephone, to solicit feedback. Interactive media, on the other hand, have embedded or integrated technologies that enable interaction to happen within its system. Therefore, the phrase "interactive media" as it is used in this book describes media where a computer interface is integral to the medium. The presence of this computer interface differentiates "interactive media" from "traditional media." To be clear, a computer interface is not limited to a personal computer; it includes wireless handheld devices, smartphones, digital two-way cable television boxes, and video game consoles to name several examples.

As the newspaper industry sought ways to deploy "interactive media," in both the pre-Internet and post-Internet periods, its leaders were keenly aware of the importance of the underlying network provided by the telecommunications industry. While newspaper publishers believed they had relevant content, they understood that ownership of a press was irrelevant for electronic distribution. The newspaper industry, which had always relied on its own distribution infrastructure from product creation to delivery, found that deploying interactive media required the use of infrastructure outside of its immediate control. Newspapers had to participate in an ecosystem created through interlocking business relationships among content providers, computer manufacturers, and network providers.

Media Convergence

Conceptually, the process of creating such an ecosystem was known as "media convergence." Media convergence is also described as a melding, or marriage, of existing forms of media and communications technology. There is wide acceptance among scholars of the premise that convergence happens due to a shift in technology from analog delivery to digital delivery. By that association, the discussion of "media convergence," "new media," and "interactive media" has been inextricably linked. Hartley wrote, for example, that "it is possible to identify the late twentieth century as an era passing from analogue to digital. . . . Even forms of larceny shifted from 'analogue' (stealing books or magazines from retailers, for instance) to 'digital' (downloading music or pictures via Napster. . .).[82] The most common concept of media convergence is that as content becomes available in digital form, devices that deliver the digital content such as television, computers, and telephony will morph into a common platform."[83]

Dominick sorted media convergence into three distinct types: corporate convergence, operational convergence, and device convergence. He wrote that corporate convergence expresses "a vision of one company delivering every service imaginable,"[84] while operational convergence describes what happens "when owners of several media properties in one market combine their separate operations into a single effort."[85] Dominick defined device convergence as the result of "combining the functions of two or three devices into one mechanism."[86] His notion of device convergence is closely aligned with other definitions that emphasize distribution methods as a major consideration. Dominick noted, for example, that "all media seem to be converging on the Internet as a major channel of distribution."[87] Dominick's three categories of convergence provide useful shorthand for discussing how the newspaper industry has viewed convergence. The industry's fixation on operational convergence, for example, receives considerable focus in this book.

Structure of the Book

This book begins by exploring the start of the online era and how newspaper companies reacted to it. It recounts the creation of the first online newspapers because it is important to consider this earlier perspective and the effects this history had on the newspaper industry as it confronted significant changes in its market. The structure of the book allows for the unfolding events to establish historical context for how and why the newspaper industry reacted as it did to the emergence of the Internet and how those experiences shape the decisions industry leaders must make today. The first four chapters explore the newspaper industry's online experiences prior to the Internet. Chapters 5 and 6 focus on the emergence of the Internet and its immediate aftermath, while Chapter 7 and the Conclusion examine the state of the newspaper industry and what that portends for the near future.

Chapter 1 provides examples from the newspaper industry's early forays into electronic publishing—projects like StarText, Viewtron, and Gateway—and explores the expectations and concerns associated with these initiatives. Reviewing the partnerships and the underlying technologies that gave life to these projects as well as the attitudes of consumers and industry leaders provides an understanding of the gap that developed between the hype surrounding their development and what actually happened once they were deployed.

Through the exploration of Knight-Ridder's Viewtron project and others, the book establishes a historical foundation for the newspaper industry's early role in developing interactive media. Examining the newspaper industry's decisions to shut down much of this high-profile activity provides insight regarding the influence these projects had on future industry decision making. Initially, many leaders viewed the closing of these videotext projects as a referendum on the ability of printed newspapers to compete with online media. Later, the financial losses incurred by these projects were recalled as excessive and contributed to the industry's reluctance to take risks when confronting decisions regarding new technology.

Chapter 2 also examines activity from the early 1980s, but focuses on the newspaper industry's flirtation with cable television. It is important to examine the influence a burgeoning cable television industry had on the decisions made by newspaper companies regarding technology and electronic distribution in the early 1980s, considering that newspaper companies spent more of their capital resources on cable television ventures than on any other form of emerging technology during this period. This chapter explores how the early development of cable television resonated with newspaper publishers who adopted the notion that a wire into every home could be a way to protect their interests in local markets.

As cable television expanded, however, these investments became very expensive for the newspaper industry to maintain, and cable television operators became reluctant to share channel capacity and revenue with newspaper companies. When this period of cable investment ended, the newspaper industry's involvement was not cast in the same light as the decision to close videotext projects. With cable, newspaper companies were seen as backing away from a distribution platform rather than a new full-fledged medium, which made strategic sense to the business and investment community. The newspaper industry experience with cable also fostered a sense that the competitive effects of cable television had been overstated, and it bolstered the newspaper industry's belief in its own competitive position as it decided to directly address incursions by the telecommunications industry.

Chapter 3 examines how the newspaper industry reacted to the possibility of direct electronic competition from the nation's telecommunications industry. Newspaper publishers already viewed the phone companies' Yellow Pages directories as advertising competition, but the prospect of those vast encyclopedic listings ported into an online database seemed like an unfair advantage. This chapter explains how the newspaper industry sought to derail that threat and, in doing so, provides additional insight into the newspaper industry's evolving perspective of its market and competition in the early years of online media.

The newspaper industry committed to an unprecedented political lobbying campaign to pass legislation or affect regulation designed to keep the telecommunications industry out of the local market information services. The industry's effort is recalled as a successful undertaking from a lobbying perspective, but its political victories ended up casting the newspaper business as defensive and protectionistic. Moreover, critics contended the newspaper industry was spending to wage political warfare when it should have been spending on technology research and development to prepare for a next generation of online media.

Chapter 4 explores the newspaper industry's relationship with a new breed of upstart companies that ushered in the era of proprietary online systems. Prodigy, AOL, and others presented newspapers with the opportunity to participate in the online market with relatively little capital investment. Newspaper companies could focus on creating content, while the online systems provided the platform. Although newspapers had made the phone company giants their public enemy, this chapter relates how publishers embraced these proprietary systems companies. After a decade of investing in interactive ventures with little return, newspaper companies liked the idea of sharing risk with these new companies. As the newspaper industry envisioned it, the online marketplace would largely resemble its offline world. In this market, large media companies centrally created the content, controlled its distribution, and relied on advertisers to pay for it all. Newspaper companies understood the proprietary online services model because it so closely reflected their own.

Chapter 5 recounts that as the newspaper industry began its major push into the online world with its proprietary online service partners, the Internet burst into the consumer market. The chapter explores the newspaper industry's reaction to the World Wide Web as it established the Internet as a formidable media distribution platform. The chapter examines several of the projects undertaken by the newspaper industry that illustrate its response during the period that led up to the Internet industry's financial bubble collapse in 2000. In explaining how this profound transition unfolded, this chapter looks first at the development of the Mosaic browser, which was critical to the overall acceptance of the Internet as a mainstream platform.

The focus then turns to the newspaper industry's migration to the Web, and highlights the New Century Network (NCN) initiative. The chapter examines the decision to close NCN and includes retrospective comments from participants and observers about what its failure said about the newspaper industry's ability to navigate the changing marketplace. The chapter also addresses how the industry wrestled with numerous issues involving content, structure, and business models, with a particular emphasis on the classified advertising component.

In Chapter 6, the Internet's post-bubble period is explored as a time of extreme change for the newspaper industry. The newspaper industry had believed it would lead media's digital transition and influence the process on its own terms. The emergence of the Internet challenged those assumptions, and the newspaper industry was marginalized as new companies transformed the Internet into their own version of what an online media platform should be. The chapter begins by discussing the AOL deal to acquire Time Warner in the context of other media mergers and how such activity led to a new perspective of convergence.

The chapter uses the activities of Media General in Tampa, Florida, as an example of how convergence became a pragmatic approach undertaken by traditional media organizations in response to their new environment. However, convergence activities had minimal impact on the financial performance of newspaper companies, which came under increasingly harsh investor scrutiny in the early 2000s. At the outset of the decade, newspaper companies struggled to understand the valuations investors awarded to Internet companies with no record of success. The chapter explores the Internet investment bubble for the effects it had on the newspaper business and discusses the reaction to the investment collapse. For a brief period, the sell-off of Internet investments led to a sense of vindication in the newspaper industry.

But the wild swings in the stock market notwithstanding, this period represented a turning point for the newspaper industry and its investors. The long-term economic prospects for newspaper companies were diminished by the emergence of Internet competition, and many investors wanted newspaper companies to embrace sweeping business reforms and articulate a long-term vision for economic viability. The chapter examines the decision of Knight-Ridder's management to close the business. The event is recalled as signaling the beginning of the end of the newspaper industry's long-established business model.

Chapter 7 looks back at the material and presents a summation of key themes that emerged from the online era that began in 1980. This chapter recalls the industry's relationship with online media in a systematic way so that it is more clearly understood how the industry arrived in its current state. The chapter connects the lessons of history by reviewing the key themes and illustrates how the newspaper industry's approach to online media left it vulnerable as the market shift to the Internet accelerated in the 2000s. Unpositioned to substitute revenue losses from the printed newspaper with gains in online revenue, the newspaper industry found its business model in ruins.

The rise of the Internet is often blamed for the newspaper industry's predicament, but losing the digital revolution should not have been the forgone conclusion. As the issues chronicled in this book reveal, the reality was an extremely complicated and nuanced clash of cultures that is yet to be fully resolved.

1

Videotext and the Birth of Online Newspapers

In the late 1990s, hyping the Internet was commonplace. Media critics, Wall Street analysts, and technology pundits all proclaimed the Internet as the next big thing. The punditry predicted the Internet would change everything by destroying some markets, while creating new ones. Such hyperbole, however, was not new. A decade earlier—long before anyone except maybe a few dedicated researchers even contemplated the possibilities of a commercial, consumer-oriented Internet— newspaper executives extolled the virtues of network-based technologies they were about to unleash.

Consider the words from Albert J. Gillen, president of Knight-Ridder Inc.'s online subsidiary, published in connection with the launch of the company's Viewtron project in 1983:

> Welcome to the future! Tomorrow has arrived. A historic moment in the United States is upon us. . . . How many of us were fortunate enough to be there when the first television set showed its first program, when the first radio crackled its first sounds, when the first talkie movie was shown, when the first words came through Bell's telephone, or when the first U.S. newspaper came off the press? Few of us actually can say that we were there when history was being made—until now. A new communications medium is making its commercial debut in the United States. A medium that combines space age technology with your everyday television and telephone line to bring you a new world of information and services.[1]

This commentary exemplifies the hype surrounding interactive media as its earliest forms debuted in the United States. This chapter provides examples from the newspaper industry's dozens of early forays into electronic publishing—projects such as StarText, Viewtron, and Gateway—and examines how the expectations, costs, and, in some cases, anxiety associated with these initiatives affected their ability to succeed. Reviewing the partnerships and the underlying technologies that gave life to these projects as well as the attitudes of consumers and industry leaders provides an understanding of the gap that developed between the hype and what actually happened. Through the exploration of Knight-Ridder's Viewtron project and others, this analysis establishes the historical foundation for the newspaper industry's early role in developing interactive media. Kyrish wrote that videotext, especially in the United States, "is often perceived as a major market failure that should have been easy to foresee."[2] This observation underscores the importance of examining the newspaper industry and its relationship with videotext, for it provides historical context relevant to understanding how this industry reacted and responded to the emergence of the Internet.[3]

The Videotext Market

When Knight-Ridder's Viewdata Corporation formally launched Viewtron in 1983 after two years of field testing, the service was hailed as "the first commercial videotex service in America."[4] Leveraging technology developed in Europe and Canada, Viewdata sought to bring interactive media to U.S. consumers by deploying videotext.[5] Teletext represented another method available at that time for transmitting and displaying text onto a television screen, but teletext was regarded as a passive technology. An audience could read teletext, but the engagement experience was limited. The excitement associated with videotext, by comparison, stemmed from its connection to a telephone, which allowed a system to serve specific content requested by a user.[6]

Videotext represented the first real melding of media (newspapers and television) and communications (the telephone) for the consumer market. Combining media forms in this manner is associated with the concept of convergence, which was explained briefly in the Introduction. While the term "convergence" is rarely applied in historical discussions of videotext, the concept it represents was important in the early adoption of videotext.

Proponents recognized the potential for linking media content to consumers through a telecommunications network. Kyrish wrote that "Expectations for videotex were strongly driven by the fact that it was technologically possible."[7] She added:

> Although the architecture of videotex may now seem simplistic, the primary sources from the time show that the technology was as fascinating and cutting-edge as today's technologies appear to us. "Convergence" is only a new term, not a new concept: articles and books about videotex positioned it as the natural result of combining the television, the telephone and computing power into a new and powerful alloy.[8]

Kyrish reported that several independent research firms made very positive forecasts regarding the future of videotext. The Institute of the Future said in 1982 that videotext would be in 30 to 40 percent of U.S. households by 2000. Strategic Inc. projected in 1981 that the consumer videotext market by 1990 would be worth $19 billion in equipment sales supported by another $16 billion in additional spending, while Booz-Allen forecasted the market would be worth $30 billion in revenue by the mid-1990s.[9] International Resource Development said in 1981 that "electronic newspapers"—a sub-market within the universe of videotext offerings—would generate $500 million in annual revenue by 1990.[10]

Such huge financial numbers emanating from prestigious market research firms contributed to the strategic planning undertaken by newspapers. Nevertheless, those revenue forecasts were tempered by the results of other studies conducted in the early 1980s in an attempt to gauge consumer receptiveness to online technologies and their relationship to newspaper consumption. One study conducted by the *Register & Tribune* in Des Moines, Iowa, found that only one-fourth of the 1,022 respondents from its market were interested in such a service, while 59 percent expressed no interest at all.[11] A study by University of Florida researchers in March 1982 asked consumers in the Gainesville, Florida, market if they "would stop buying a newspaper if they could get its contents on a TV screen?" Sixty-five percent of the 373 respondents said "no."[12] Studies such as these confirmed for the industry that only a small portion of its audience was interested in online media services in the early 1980s. Newspaper companies moved forward with investments in videotext projects even though they believed there was little consumer demand for such services in these early years. In the absence of perceived consumer demand, there were other factors that motivated newspaper companies to enter the market for videotext services.

Competition Effect

Udell wrote in 1978 that the "so-called experts have predicted the demise of newspapers as we know them since the 1930s."[13] Radio and later television presented formidable competitive threats, but after surviving—and thriving—against the electronic media throughout the 1950s, 1960s, and 1970s, the newspaper industry had no reason to believe it could not coexist with computer-based competition. For the most part, industry leaders were confident that a significant competitive threat to its core printed newspaper products from computer-based media remained in the distant future. But this outward air of industry confidence was buffeted by a nagging suspicion that such thinking could be wrong—that doing nothing would leave the newspaper industry vulnerable to competitors that acted unlike other electronic media.

Computers of earlier generations were understood to be text-based devices, which allowed them to be closely associated with print competition. There was concern that computers connected to a content provider 24 hours a day would upset the traditional news cycle, which had favored morning newspapers.

Kyrish maintained that it was the need to keep up with potential competitors that motivated the spending on early videotext endeavors by Knight-Ridder and other newspaper companies. She wrote that such "spending was based on concern that new technologies would reorganize existing structures and that companies that did not invest would be trampled."[14] These projects were, in essence, the result of risk management rather than a strategic desire to be innovative. In the early 1980s, newspaper companies were concerned about the expanding market for cable television. Publishers were aware that cable television represented a network conduit into the home that someday could offer an interactive experience. Warner Cable had attracted significant attention in the late 1970s when it introduced its own version of an interactive system in Columbus, Ohio, called QUBE.[15] While some critics viewed it as "a novelty," others saw it as "a major innovation in marketing research and segmented communications," which "offers substantial advantages for added flexibility in programming and advertising."[16]

Newspaper executives felt they had no choice but to pay attention to the burgeoning market for cable television and the medium's potential for ancillary services. How newspapers became intertwined in the cable television industry is explored in greater detail in Chapter 2. The purpose of this chapter is to explore the newspaper industry's involvement in videotext in the early 1980s and the factors influencing those projects. As explained in the Introduction, the early 1980s was a time of much rhetoric regarding the Information Society and the coming information-based upheaval in industry that would rival the Industrial Revolution. As such, the newspaper industry pursued videotext projects not only because of concerns over competition, but also from a desire to be seen as a leader in the information revolution.

Technological Determinism

The early 1980s marked the emergence of the computer as a consumer tool and provided the impetus for much of the concern about the future of media—especially in printed form. By 1983, the personal computer had made such an impact on society that *Time* magazine recognized the device as its "Machine of the Year," which replaced its usual "Man of the Year" feature.[17] In looking back over the evolution of media, Bagdikian wrote that "the history and subsequent emergence of the computer into the modern media scene is as significant as the invention of high-speed presses was to the history and social effects of newspapers and magazines."[18] Therefore, it is important to examine how videotext came to represent, especially for the newspaper industry, the method in which the promises of computer technology would be fulfilled. As the newspaper industry became enamored with the role it could play in bringing new technologies to market, many in the industry viewed the development of online media as inevitable; that newspaper companies had no alternative other than to participate. Flichy described this as succumbing to a "totally deterministic perception of technology."[19] Boczkowski wrote that the development of videotext technology within the cultural climate of the early 1980s allowed for it to become symbolic of the "'information society' rhetoric [that] was popular in both the press and scholarly works."[20] He noted that the culture was primed to accept "these new technologies" as "symbol of an epochal change. . . ."[21]

Clarifying Terms

Knight-Ridder's Viewdata Corporation traced the origin of videotext to Sam Fedida, a British researcher, whose work in the early 1970s led to the British Post Office's deployment of a service known as Prestel in 1979.[22] As British researchers were readying their system for commercial deployment, government-backed projects were also underway in France (Antiope and later Teletel), Canada (Telidon), and Japan (CAPTAIN).[23] While the computer engineering in these systems may not be identical, they are universally discussed as videotext systems.

The phrase "viewdata" was used interchangeably with the term "videotext" in the late 1970s and early 1980s,[24] but "videotext" will serve in this book as the name for the generic system, to avoid confusion due to Knight-Ridder's adoption of "viewdata" as the corporate name of its videotext subsidiary, Viewdata Corporation. Sigel's definitions, published in 1983 and often cited in the literature, provide the foundation for understanding the terms as they are used in this book. Sigel defined videotext as "a means of displaying words, numbers and pictures on a TV screen at the touch of the button." Sigel explained that teletext—another platform deployed in the early 1980s—was "a broadcast system [that] involves the one-way sending of pages of information"; videotext "involves the sending of information from a central computer to an individual terminal over telephone lines."[25]

Teletext was essentially a passive broadcast medium, offering its user nothing more than the ability to read text as it scrolled over a television screen. Videotext, however, offered a connection to a central server—a communication path to talk with the content provider and interact with the content in new ways. This concept of interactivity, as explained in the Introduction, was inherent in how Sigel differentiated the two technologies. He wrote that teletext is one way and offers no path for the user to select content, adding that although it affords the user the ability to "determine the timing of what he sees, teletext is not a truly interactive medium." Videotext was the term that became primarily associated with early online systems that allowed for "both the selection of information and the timing of its display [to be] determined by the recipient."[26]

Neustadt explained videotext in much the same way as Sigel. Neustadt described videotext as two way, adding that in a videotext system: "The computer holds a large number of pages (a database), and the user sends a signal to it to request the desired page. The computer then transmits that particular page. Videotext sends different pages to different users and can handle multiple requests simultaneously." He also defined teletext as a "one-way system, with signals flowing to the user."[27]

Aumente also presented interactivity as the key attribute of videotext. His basic definition is similar to the others presented: "various computer-based interactive systems that electronically deliver screen text, numbers, and graphics via the telephone or two-way cable for display on a television set or video monitor."[28] He explained that the interactive aspect of videotext involved multiple relationships: communicating with the system provider and others using the system as well as transacting business with service providers such as banks and airlines.[29] He wrote: "Interactivity catapults existing media habits into an entirely new realm."[30] Because of this interactivity, the potential for videotext as a new communication medium was exciting to its early devotees. It was recognized as enabling a paradigm shift.

An Extension of Electronic Publishing

Early experiments with videotext captured the imagination of researchers and scientists who envisioned the power of computer-based information systems as a replacement for print-delivery systems. As Sigel wrote: "It is undeniable that video-text has many advantages over print, among them speed, selectivity, personalization of information and the maintaining of wide-ranging, comprehensive collections of data."[31] Neustadt espoused that the technology developments were leading to "a new mass medium" he described as "electronic publishing."[32] He wrote:

> Until recently, mass distribution of information has been dominated by publish-ing and broadcasting. Now, technology is marrying these media to spawn a new one: electronic publishing. Print-type information—text and graphics—is being distributed over electronic channels: television, radio, cable TV and telephone wires. In the past four years, electronic publishing has changed from futuristic fantasy into a serious business.[33]

In this explanation, electronic publishing is associated with electronic deliv-ery, but this meaning evolved out of computer-based production processes. Researchers in the Internet era are likely to take the phrase "electronic publish-ing" for granted, but it was new in the late 1970s and represented an emerging construct for the publishing industry, especially newspapers. The earliest use of "electronic publishing" was attributed to a conference held in the spring of 1977 by the U.S. Institute of Graphic Communication.[34]

Around that time, the phrase was generally understood to encompass two meanings: first, the use of computers to facilitate production of a printed product through photocomposition; and second, the use of computers and telecommu-nications systems to distribute data to users electronically.[35] Lerner argued that electronic publishing represented a holistic concept that applied to both produc-tion and distribution of information.[36] Other researchers such as Cuadra[37] and Gurnsey[38] echoed Lerner, concurring that the concept applied to electronic sys-tems used to produce printed material as well as material distributed online. All of these early attempts to define electronic publishing illustrate the newness of computing in the publishing industry at that time.

Even before the phrase "electronic publishing" became an industry-recognized term, the U.S. newspaper industry began deploying computer systems to improve the efficiency of its front-end production processes. These new systems included technology to scan text written on typewriters into computer production systems and later full-text entry systems with video display terminals (VDT).[39] Word processing software and computer-based publishing systems are commonplace today, but they represented sweeping fundamental change when introduced. To understand how radical these systems must have seemed to the reporters and edi-tors who were among the first to use them, consider Udell's description:

The reporter can read what he or she has written and can change it on the keyboard and screen as often as wished. When finally satisfied with a story, the reporter simply informs the appropriate editor who can "call up" the story from the computer onto his own VDT screen and edit it as necessary. When all editing is completed, the story remains in electronic storage to be called up whenever wanted for virtually instantaneous electronic typesetting.[40]

As newspapers came to rely on computers in the production process, an interesting phenomenon occurred. The industry, long in the technology shadow of television and radio, experienced a sense of competitive resurgence against its electronic competitors as the 1980s began. Given that computers were so new at assisting with editorial tasks, newspaper executives marveled at their use. Indeed, many believed computers assisting with production processes signaled that newspapers had crossed the line into a form of electronic media, somewhat equivalent technologically with their television rivals. Compaine wrote: "The technology has brought newspapers into the electronic age, if not with the same immediacy as television or radio, then with many of their techniques for instantaneous and remote transmission of the news back to the waiting pressroom."[41] Adopting computer technology in this manner seemed to provide the newspaper industry with a sense of new-age sophistication that had been lacking. Using computers to improve production process and enhance efficiency represented a break from the industry's industrial, manufacturing roots and afforded the opportunity to be viewed as contemporary and relevant. The initiatives to integrate computer technology into the production process were deemed successful as Compaine further observed:

> The new technology has breathed new life into this oldest of news media. Its speed, accuracy, and flexibility have helped newspapers hold down costs, brighten the product, and, in some cases, improve the editorial package.[42]

In this context, it is easier to understand why newspaper companies were quick to adopt videotext. Online distribution was an extension of the electronic publishing efforts already underway, and videotext was considered a natural extension of the computer-based tools they were using. Picard wrote: "Because the news processes associated with the technology captured keystrokes, it was now possible to reuse or easily alter content prepared for the newspaper for use in a videotext operation."[43] As such, electronic publishing evolved as a blending of the production process with the delivery process in such a way that a new media form emerged. It was, as Kist observed, a significant advancement, "not merely one more development along the continuum in the widespread dissemination of knowledge which began with the invention of moveable type."[44]

Kist also recognized that electronic publishing was not an invention in its own right, but an amalgamation of inventions that included such things as computers, telephony, and photography. With that construct in mind, he wrote that electronic publishing represents:

> . . . more than the transfer of characters to a screen or to a printer; it is more than faster and cheaper typesetting; it is also more than an efficient means of storing and retrieving documents. [Electronic publishing] offers the possibility of bringing a vast store of information and knowledge . . . directly to the user.[45]

This view of electronic publishing reflects the emergence of convergence concepts that will be explored further in subsequent chapters. At this point, however, it will suffice to understand that videotext became a pragmatic tool for those seeking a technology platform that would allow them to realize the potential of electronic publishing beyond enabling a printed form. In other words, videotext became the platform that first expanded the scope of electronic publishing.

Such themes were prevalent throughout numerous articles and books published in the 1970s and 1980s, including one that noted that electronic publishing had become a "hot topic" largely because "it has been heralded as one of the portents of the impending information age which is expected to usher us into the lap of a postindustrial era."[46] *Editor & Publisher*, a leading newspaper industry trade journal, also called the transformation of its industry's production processes a "technological revolution."[47] Even though such phraseology was used frequently when describing the Information Society at large, it was particularly pertinent when used within the newspaper industry. As Boczkowski explained, the specific use of such phraseology within the industry setting underscores its influence. He wrote:

> This "revolutionary" language points to an ideological trait that also contributed to create a context conducive for the appropriation of videotex by American dailies: the technologically deterministic belief that electronic publishing would drive the future of the industry.[48]

The rise of electronic publishing and its association with Information Society rhetoric is an important element to consider when exploring the factors that motivated newspaper companies to take the actions they did in the early 1980s.

The Influence of Rhetoric

As videotext was deployed through commercial services aimed at consumers, it came to be seen as the tangible manifestation of the Information Society rhetoric. Many authors from industry and academia tended to approach electronic publishing and the videotext tool from a technology utopian perspective; "revolution" was a commonly used word. The following three examples illustrate this point.

Smith believed that in terms of human communication, writing and printing represented two revolutions, and suggested the electronic publishing era constituted a third revolution. He wrote:

> Today the computer, which was developed originally as a device for calculating, has now become a device for handling text in many forms, and this interconnection between computer and text is coming to exercise so transforming an influence upon the human institutions that adapt to it that one may justifiably consider whether a *third* great turning point in information systems has come about.[49]

Weingarten, in an article based on testimony to the United States Congress, summarized the climate of discussion this way:

> It has become common in the press and popular literature to speak about the new "Information Society" or "Information Age." Whether or not such statements suffer from journalistic exaggeration, we are clearly in the middle of fundamental transformation of the way information is created, stored, transmitted and used.[50]

Alber, though generally conservative and dispassionate in his work on videotext and teletext, also opened his book by placing the subject within this heady, technology-revolution context:

> We live in a fast-paced and rapidly changing world, a time in which writers try to capture the essence of what is happening with phrases like *The Wired Society*, *The Third Wave*, and *Megatrends*. Videotex and teletext are the electronic children of this age. By 1995, they will have grown up to be an integral part of our lives.[51]

Although popularized in the early 1980s by such commentary, the concepts that made up the construct of an Information Society were not new. They had been widely discussed in the 1970s, and the origin of the core concepts is even earlier. Beniger reported that economist Fritz Machlup, working in the 1950s, advanced the concept of information as the researcher "who first measured that sector of the U.S. economy associated with what he called 'the production and distribution of knowledge.'"[52] In the late 1950s, information concepts became more closely associated with computers. How these devices would allow for information to be processed more efficiently became an important topic of that time. An article by Leavitt and Whisler in a 1958 issue of *Harvard Business Review*,[53] for example, represents one of the earliest documented uses of the phrase "information technology."[54] Subsequently, the transformation of information from an analog to digital form has been described as a key attribute of information technology.[55] For traditional media companies—especially those producing a printed product—adopting information technology implied an acceptance of a new digitally based future. In the case of newspapers, this required purging analog-based production processes often rooted in decades—if not centuries—of tradition. It was no small undertaking.

Bagdikian was among the early explorers of how this transition from analog to digital would apply to the media. He became very influential by expressing his Information Society views and how traditional media, especially newspapers, would be affected by electronic publishing. Bagdikian's prescient work, published in 1971, discussed electronic media and computing and the potential for technology to change what he considered a print-centric media landscape.

Bagdikian's work did not achieve the level of popular notoriety of the more futurist-type works by authors such as Toffler[56] and Naisbitt,[57] but his contribution to explaining the effects of technology on media cannot be understated. Bagdikian explored many of the concepts that would later inform the definitions of electronic publishing and contribute to the work by other authors and researchers. Bagdikian drew from the commentary of McLuhan, who had espoused the notion that that "printed words are an invention contrary to the inherent nature of man."[58] Bagdikian expanded on this idea:

> The new electronic media represent a return to a richer and more natural way for man to participate in his environment, engaging more of the senses and more levels of the brain than those used for abstract reasoning. As new generations respond to this multisensory medium, there will be a revival of the dominance of preprint communications—sight, sound, smell, touch, taste—and a disappearance of the "tyranny of print."[59]

Bagdikian stopped well short of agreeing with McLuhan's assertion that printed media was a dying artifact—in fact, arguing just the opposite.[60] But Bagdikian's work is seminal, for it acknowledged the influence of such ideas. He contributed to an understanding of the historical context in which decisions were made by traditional media including newspaper companies. Bagdikian wrote: "The McLuhan influence has been more dramatic and, in our time, more influential creating not only a popular dogma, but a significant body of belief among some scientists, scholars, academics, and operators of the mass media."[61] Although Bagdikian did not believe that print would die as a result of new technology, he was convinced that computing would alter the way printed media would function:

> In the long run, the more powerful substitute for print will be the routine storage of information in computers. . . . The computer can store enormous quantities of information . . . if these words, ideas, subject references, paragraphs and whole documents are indexed and coded as they are introduced into the computer's memory, the memory can be searched for particular parts of its content and they can be extracted quickly.[62]

And, in what may pass as one of the earliest descriptions of what we now know as a search engine, Bagdikian wrote about the coming power of information on demand:

> The human reader may ask the machine to use its enormous speed to search its memory for those items the reader is interested in and present only selected information.[63]

In other words, Bagdikian recognized that the consumer would have new control over information, and would be newly empowered because of that control. By the end of the 1970s, traditional media companies—especially newspaper publishers—had generally accepted that idea. Newspaper companies understood that technology would continue to alter the media marketplace even though the personal computer was a hobbyist activity. Newspapers had arrived at this understanding, in part, due to the success of electronic information services sold to businesses. How these commercial services influenced developments in the media marketplace is the topic of the following section.

The Influence of Commercial Services

Large electronic databases of information targeted at businesses and libraries were commonplace by the early 1980s. Mowshowitz observed that computer databases of information had been in existence for years, noting that services such as Medlars developed by the National Library of Medicine traced its origin to the mid-1960s.[64] He wrote that there were "thousands of databases accessible through various computer networks,"[65] including Dialog Information Services, a commercial system that contained more than 35 million records.[66] Services such as Orbit and Dialog provide examples of commercial databases that carved out market niches for specific types of information, and in doing so, established an electronic information marketplace.[67]

Online magazine described the year 1977 as "arguably the dawn of the commercial online world,"[68] but the article noted that at the time users were "limited to people directly involved in academia, defense, government research agencies, computer systems development, and a few brave customers scattered among leading organizations."[69] Nevertheless, the growth of these commercial services in the late 1970s and early 1980s was significant. The number of such services increased to more than 1,000 by 1983 from only 300 in 1975.[70]

The role these commercial database companies played in influencing the marketplace was exceptional. They proved that a marketplace existed for text-based information and data in the absence of a printed form. As researchers at the time studied the commercial market, they began to consider the possibilities of similar information depositories becoming available to consumers. In doing so, videotext was heralded as the way to deliver the same vast amounts of information to consumers at home in a cost-effective way that had not been accomplished.

Mowshowitz observed that the significant social shift occurring in the early 1980s was the ability to deliver information databases to consumers at home.[71] Through videotext, the technology was deemed to be in place to deliver the connected, integrated electronic experience required of an Information Society. Mowshowitz wrote:

> The social significance of computer-communications is hard to exaggerate. In the near future most homes will be equipped with computer terminals linked by telephone, cable, satellite or other means to regional and national networks offering a variety of databases and information processing services. A decade ago this projection might have been dismissed as visionary; now the only questions are how fast terminals will be introduced and what factors will determine the rate of diffusion.[72]

One example from this period illustrates how some commercial services began to bridge the gap between the business and consumer markets. CompuServe was created by H&R Block in the early 1970s to provide computer time-sharing services to companies that needed such resources, but did not want to invest the capital in systems of their own. As its business grew to more than 700 clients, including 120 of the Fortune 500, CompuServe began to add data services such as financial information to its product offering.[73] When personal comput-ers entered the consumer market, CompuServe decided to market its services to users who were initially identified as computer hobbyists. In 1980, CompuServe charged these consumers $22.50 per hour to access its system during the day, but only $5 per hour after 6 p.m. when most of its business clients had closed and the system had excess capacity. CompuServe provided consumers with an array of information and services, including stock quotes, commodity news, electronic mail, a bulletin board service, tests of electronic games that could be played between computer users, and even content from *Better Homes & Gardens* magazine.[74]

CompuServe's most formidable competition in the consumer market did not come from other commercial services or the newspaper industry, but rather from the magazine industry. *Reader's Digest* created a service known as The Source, which offered similar information services. Additionally, both CompuServe and The Source appealed to home computer users as a marketplace for downloading software.[75] By 1985, CompuServe and The Source were serving 325,000 subscrib-ers.[76] Although this number reflected a core audience among the growing number of home computer users, it was too small to impress the owners of the nation's large regional daily newspapers. Most of the regional daily newspapers had more subscribers than that in a single market.

Meanwhile, the commercial services market exerted greater influence on the newspaper business following 1980, when services marketed as electronic libraries began to proliferate. Many newspaper companies discovered in the early 1980s that it was easy to provide access to electronic content archives given that the raw material had been converted into digital form for the computer production processes that were used.

Dow Jones launched its Dow Jones News/Retrieval system in 1980, provid-ing online access to its Dow Jones newswire and selected content from its finan-cial newspaper, the *Wall Street Journal.*[77] That same year, the *New York Times* began offering access to its stories in full-text form through a service it called Infobank.[78] By 1985, about 50 newspapers were making full-text versions of their content available either directly through their own services or through a number of third-party electronic library services offered by Nexis, Vu/Text, Dialog, and Data Times.[79]

The electronic library services were targeted at information professionals, especially research librarians, but newspaper companies were interested in using the platforms to learn more about managing electronic information. Donald Wright, president and chief operating officer of the *Los Angeles Times*, wrote:

"With electronic libraries that are coming into use, we are developing knowledge in data-base management."[80] That sentiment summarizes the influence commercial databases and electronic libraries had on the newspaper industry in the early 1980s. Many newspaper companies came to see their operations within this context—keepers of vast databases of information that could be marketed as ancillary services.

But an industry consensus failed to emerge regarding how to translate experience with electronic libraries into a business model that held long-term promise for selling content online. Knight-Ridder, for example, seemed buoyed by the prospects of electronic libraries and continued to invest in the business at the same time it was rolling out its consumer-focused Viewtron service. However, the New York Times Co. in 1983 sold its Infobank service to Mead Data Central, then the parent company of the Nexis database.[81]

This opposite approach to one facet of the electronic information market by two of the industry's most prominent companies underscored the uncertainty of the time. It illustrated how newspaper executives at different companies were looking at the same underlying strategic issues, but arriving at different actions. This chapter next examines several factors that were important to newspaper executives as they sought business models that would support electronic distribution of their content.

Business Considerations

Even though some of the shrill rhetoric of the time predicted the imminent demise of printed newspapers, few, if any, newspaper industry leaders believed it. Some expressed concern that videotext or other forms of electronic delivery posed a short-term competitive threat, but any doomsday scenario was couched in terms of far-off years. A special trade publication report, for example, offered one prognostication that "newspapers, as we know them in 1980, will be non-existent in 2030."[82] A 50-year horizon was no death sentence at all; the leaders at that time would all be retired by 2030, and they had confidence that leadership emerging from the next generation would be equipped to deal with industry changes as they evolved. In searching for a business model, therefore, the threat to the core printed product became a secondary concern. The issue at hand was determining how videotext services could provide a return on the investment required to create them.

In critically assessing the business climate of the time, talk of leading a revolution may have appealed to the ego of some industry leaders and cannot be dismissed as a factor that led to some of the projects. But newspaper companies were conservatively managed businesses, and a review of trade publication material from this period shows that newspaper companies appeared genuinely concerned about finding a business model that would support online endeavors as the market developed.

As explained earlier, newspaper executives understood that several companies had found success selling information databases—including news services—to businesses and libraries, but they were uncertain if that model could be translated into the consumer market. Consider the comments from J. Christopher Burns, vice president/planning for the Washington Post Co.:

> We have to distinguish between the consumer market and the business market. The business market is real and it is growing. Businesses can convert the news to profit if the news product is timely, relevant and accurate. So, up to a point, the more appropriate the news they buy, the more money they can make. . . . [The consumer] is not likely to buy more news, even more specialized news, if there is only cost and no benefit. He would buy it if he absolutely needed to know more. But it is not clear that he wants to know more.[83]

Newspaper executives considered business models that would expand their reach beyond news and information; they wanted to accommodate interactivity and offer interactive shopping services and online banking.[84] As these discussions moved from the theoretical possibilities of online technology to deploying actual experiments, newspaper companies embraced videotext and ordained the technology as the platform that would deliver an Information Society and all that entails. From a business model perspective, the decision to deploy videotext was rooted in risk management. Newspaper executives had no appetite for spending to develop their own technology when videotext was available and had been deemed an acceptable platform by numerous experts.

Neustadt and Alber, for example, studied videotext holistically and considered its functionality to be extremely versatile. They considered videotext capable of delivering an array of applications that are more in line with what we think of today as e-business or e-commerce solutions. Alber wrote that "videotex applications may be grouped into six classes: information retrieval, commercial transactions, electronic messaging, educational services and personal transactions, computations and gaming and teleservices."[85] That videotext could do all of these things appealed to newspaper managers who believed a common platform would be easier to manage, but in practice, newspaper companies were most interested in the information retrieval aspect of the technology. They wanted to know whether videotext represented the next wave in development toward a paperless society.

As computers became more closely associated with information processing, speculation increased about how they would transform print-based publishing. Lancaster, for example, wrote that the promise of computers and their abilities in the manipulation of text led many in the 1970s to consider the possibilities of "paperless information systems."[86] He noted that increasing productivity concerns led the National Science Foundation in 1975 to recognize the "need for a replacement for paper."[87] Lancaster concluded that the shift from print-based communication to electronic-based communication was inevitable because of the rapidly increasing amounts of information and the human labor required for producing and distributing it. He wrote: "Computer processing offers the only possibility for coping with the situation because substantial improvements in manual productivity are infeasible."[88] Newspaper executives seemed eager to discern how far down the paperless path they could travel by deploying videotext. Newspaper companies did not believe that turning on a videotext system one day would mean shutting down the presses the next day, but the possibilities of electronic distribution were exhilarating given what had taken place regarding newsprint costs in the 1970s.

Publishers were dealing with steep price increases in newsprint—the basic raw commodity needed for production. "After years of plentiful supply and almost constant price, newsprint demand exceeded supply and higher costs for pulp and energy forced a rapid escalation in newsprint prices between 1973 and 1979," wrote Compaine.[89] To manage costs, publishers cut the size of their papers and implemented strict waste controls in the pressroom. Faced with the increasing newsprint expenses as well as increases affecting labor and delivery, newspaper executives seemed attracted to videotext as a technology that would help them reduce production costs.[90]

When considered in the historical context, the idea of replacing paper entered into the strategic planning discussions taking place within the newspaper industry regarding electronic publishing and distribution. An almost romantic notion emerged that new technologies would not only save on commodities, but also cure some of the shortcomings associated with the print products by allowing them to be updated continuously and by reporting the stories unconstrained by space limitations.[91]

By adopting videotext, the newspaper industry would be part of creating a medium distinct from other media, one that deployed databases on demand and interactive services melded to create a new kind of information experience. In this new world, newspaper executives envisioned information delivery not based on a story or a narrative found in television, radio, or print, but on something akin to the encyclopedia. It would be a syntactic resource not linked to temporality, but available for constant exploration. Radio and television had added sound and motion to the story form, but it remained linear, passive, and temporary. The new form of delivery would be interactive, passing an element of control to the user and thereby empowering the individual.

This concept of user control and empowerment, which were at the core of Information Society rhetoric, began to make its way into business model discussions as well. Newspaper executives, for example, wanted to know whether users were willing to pay a premium for interactive services such as message boards and electronic mail. They wanted to know whether certain types of content would be worth more than other types of content.

As the newspaper industry considered how it would deploy new online technologies, these types of concerns involving the financial ramifications of content offerings were always an undercurrent in the discussions. Some observers, however, believed that editorial concerns had taken over the debate.

Russell, an advertising scholar, observed that much of the traditional media industry's angst about providing electronic distribution focused on content and societal concerns surrounding a free press, rather than financial matters. "Lost in much of the debate is the long term effect on financial support of the media and, of course, the future role of advertising," he wrote in 1978.[92] Russell understood, even then, that interactive technology would inherently change the way advertising worked by placing the end-user in a position of having more control over the media experience.

> The role of advertising in such a system largely remains to be worked out. However, it is obvious that the system differs in fundamental ways from present advertising delivery through the mass media. Here it is the consumer that controls the type of information ordered rather than having a potpourri of advertising included in the mass media. Consequently, the advertiser has the advantage of dealing only with prospects, but has fewer opportunities to gain new customers who may be unfamiliar with a specific product or brand. Such a system also blurs the differences between advertising and editorial information. For instance, if the airlines underwrite a service to provide airline schedules . . . is this advertising or not?[93]

When advertising issues did bubble into the public discussion, the industry appeared to be most interested in how to best protect its classified advertising franchise. This type of advertising was all-text and organized by product categories, which made it vulnerable to electronic competitors who could put the advertising into a database and make it searchable. Newspaper industry leaders understood that computers connected to databases of classified advertising could threaten their long-held classified advertising franchise—text-based advertising that formed the industry's economic foundation. Heretofore, newspapers had shared very little of the classified advertising pie with its radio or television competitors. As Compaine noted: "Classified advertising . . . does lend itself more to these futuristic delivery modes," and he warned "this would appear to be one area in which newspapers may well have to take the lead, before . . . others usurp this function."[94]

The Newspaper Advertising Bureau in October 1980 launched a task force called "Classified and the New Technology" to investigate the technical requirements for offering classified advertising online as well as explore competitive threats from companies outside the newspaper industry.[95] Kauffman wrote that the potential existed to "make the electronic database into an extension of the newspaper," but he added that the technology "can also whet the appetite of the broadcast media—or others—for a share of the attractive classified advertising business."[96]

Examples from the early 1980s illustrate that advertisers and their agencies approached the new online platforms as an opportunity to marry content and advertising more closely than had been accomplished with other media through sponsorship models or even through product placement.

Young & Rubicam, in one early videotext-based campaign, created an online tea catalogue for its Lipton client. It combined the catalogue with editorial content such as serving suggestions and recipes and distributed the package to Knight-Ridder's Viewtron system, which had 1,600 users at the time.[97] The agency also developed an online guide for baby care on behalf of client Johnson & Johnson and one on home repair for Olympic Stain. By the end of 1983, Young & Rubicam had nine full-time employees engaged with experimental videotext systems. In discussing the importance of integrating advertising closely with content in the new medium, an executive with the agency said: "The most effective way to sell products is to get the consumer involved."[98]

Ogilvy & Mather is an example of another prominent advertising agency that was producing online specialty content for advertisers to distribute on upstart videotext systems. Executives with both Ogilvy & Mather and Young & Rubicam were open about the experimental nature of their endeavors, but once again there was a sense that they were responding to forces outside of their control. As the director of new electronic media at Young & Rubicam stated: "We're operating on the premise that this is inevitable. We don't know where or when, but it's inevitable."[99]

In reflecting upon the myriad list of influences and business considerations that led newspaper companies to embark on electronic distribution using videotext, Boczkowski concluded that no one factor can be singled out above another. He asserted that it was the combination of several cultural and economic factors that led the newspaper industry on its quest to deliver videotext services.[100]

To make sense out of these influences and to understand why they were influences at all, however, requires an understanding of the industry's economic context—the business climate in which newspapers operated during the late 1970s and early 1980s. The details of the industry's economic circumstances were provided in the Introduction, but it is important to recall one salient aspect of that discussion. Although the newspaper industry was financially robust during this period, many leaders expressed a sense of vulnerability that could be traced to the industry's failure to increase its audience relative to the nation's population growth. The newspaper industry was losing market share, and it was an irksome trend.

Udell's work, generally an optimistic text on the newspaper industry's prospects, acknowledged that "circulation growth has lagged the expansion of population."[101] The declining penetration rates represented a potential long-term problem in terms of retaining and attracting advertisers. If newspapers were losing audience relative to population, advertisers could assume that newspapers would also lose audience share relative to competitive media. And erosion in advertising dollars would not bode well for the industry's long-term trend given that its costs were rising significantly.

That context, according to Boczkowski, makes it easier to understand how electronic delivery emerged as such an important issue in the early 1980s, even though newspapers were very profitable and did not require any immediate relief. Boczkowski wrote that the industry at the time became focused on electronic distribution as an opportunity to remake itself so that it could improve the industry's long-term viability.[102] He summarized the business climate this way:

> . . . there was a perception among many analysts and actors that trends such as decreasing penetration, rising newsprint and distribution costs, readers moving to the suburbs and getting news on the radio while driving to work, less homogenized consumer tastes challenging mass advertising, and difficulties in attracting and retaining younger readers seemed to compromise the long-term viability of print as an information platform.[103]

Boczkowski's succinct summary explains the newspaper industry's collective motivation. The long-term macroeconomic concerns represented huge problems for an industry that had no immediate solutions to them. How these issues were addressed by the various newspaper companies influenced the business models that were deployed in relation to their online activities. For example, companies that emphasized cost control tended to take a "wait and see" approach to electronic distribution and videotext. Some of these more cost-influenced companies adopted projects based on low-cost bulletin board systems. Companies that were more concerned by broad competitive implications or intrigued by the strategic possibilities tended to invest more aggressively in electronic distribution. In the following section, this chapter explores examples of several types of newspaper videotext initiatives.

Newspaper Videotext Initiatives

In Europe and Japan, videotext was largely the purview of government-sponsored initiatives. In the United States, Boczkowski contended that—in the absence of government subsidies—videotext floundered until the newspaper industry stepped forward and took a leadership role in its development.[104] Boczkowski wrote that "a mix of a changing economic environment, a massive computerization of the industry, and a technologically deterministic ideology paved the way for this interest in consumer-oriented videotext by American dailies."[105] This statement underscores the factors discussed in the previous section as motivation for the newspaper industry taking action with videotext.

Although the newspaper industry emerged as a leader in videotext development, this industry was not alone in the early market development. Many others were represented in the early marketplace, including television, magazines, commercial database vendors, and an array of technology companies. This mix of entities—including newspapers—created a trade association known as the Videotex Industry Association (VIA) to represent those involved in the market. The newspaper industry's involvement in videotext, however, stirred mixed emotions among the association's membership. Newspapers expressed interest in the market and represented a large source of capital to expand the market. But skeptical observers doubted that the newspaper industry had critically assessed the market opportunity, concluding that most newspaper executives held unrealistic expectations for what could be accomplished in the short term.

When Belo Corp., publisher of the *Dallas Morning News* in Texas, shut down an early project, leadership in the VIA called on newspaper companies to proceed with caution. They were concerned that newspapers not committed to the long-term growth of videotext would end up creating a negative image for all involved.

Larry T. Pfister, the group's chairman, pointedly told newspaper publishers in the summer of 1982 "not to get romanced or stampeded into getting into the business. If you don't have the economic staying power, you almost are going to guarantee a failure."[106]

Those words turned out to be prophetic as the newspaper industry continued its rush into videotext with numerous high-profile projects and partnerships only to retreat from the market by the middle of the decade. The experiences of individual companies within the newspaper industry varied considerably during this period. However, exploring several prominent examples provides insights into what transpired and helps to explain the aftermath. The following sections look deeper into the industry's first online initiative and are followed by a discussion of StarText, Viewtron, and Gateway.

The First Newspaper Online

The first newspaper content delivered to consumers via a computer occurred on July 1, 1980, when the *Dispatch* of Columbus, Ohio, transmitted several articles through a system created with its partners CompuServe and the Associated Press.[107] The launch culminated four months of planning and about 300 hours of programming time required to link the newspaper's computers with CompuServe's computers that were also based in Columbus.[108] The newspaper and its partners called the project an experiment and launched it with little fanfare. At the time of this experiment, there were only about 3,000 home computers in the entire country, and only about 250 of them were in Columbus and could access the newspaper's content.[109] Nevertheless, Robert M. Johnson, the newspaper's vice president and general manager, said that the *Dispatch* "had proved to itself that it can be done."[110] For that reason alone, this early experiment is seen as historically significant. Merely proving that the newspaper could link its computers to an online service and transmit the content was viewed as a major leap forward. But as soon as that milestone was achieved, the discussion quickly turned to the prospect of customer acceptance.

Rittenhouse had observed that much of the effort surrounding computer-based communication technology directed at the consumer market "assumes the demand exists with little or no justification for this claim."[111] Keith Fuller, president and general manager of the Associated Press during the *Dispatch*/CompuServe experiment, acknowledged as much: "Our board has heard two views: One that electronic delivery is the future knocking at the door, and the other that electronic delivery to the home is a disaster hunting a victim. We intend to find out which is the case."[112]

Laakaniemi referred to this issue as the "Chicken or Egg Syndrome," noting that for the project to be successful there needed to be more users with computers capable of accessing the content, but there needed to be more content to attract users.[113] Jeffrey Wilkins, president of CompuServe, viewed the prospects for the nascent market in terms of personal computer adoption, stating that it "was not a question of whether a market exists for an electronic [newspaper] edition, but when." He added: "The question is whether the computer in the home is a 1980s business or one for the year 2000."[114]

While still experimenting with the *Dispatch,* CompuServe expanded distribution to include the QUBE interactive television system, which was also operating its experiment in Columbus. Through its partnership with the Associated Press, CompuServe also signed on several other newspapers to participate.[115] The project attracted the national elite newspapers—the *Washington Post,* the *New York Times,* and the *Los Angeles Times*—and several influential regional dailies, including the *Minneapolis Star Tribune,* the *San Francisco Chronicle,* the *San Francisco Examiner,* the *Virginia-Pilot,* the *St. Louis Post-Dispatch,* and the *Atlanta Journal-Constitution.*[116]

Even though the project had strong industry participation, expectations were muted. Consider this comment by an editor at the *Atlanta Journal-Constitution:* "If response is small, it could mean the advances in electronic information processing have, for the moment, outstripped consumer demand for 21st century gadgetry."[117] Indeed, the CompuServe project with the Associated Press and its member newspapers ended in June 1982—about two years after it started—as publishers lost interest in a system where so few users accessed their content. According to Hecht, news contributed by either the Associated Press or the newspapers generated only 5 percent of the system's traffic.[118] Independent research of system users conducted by RMH Research on behalf of the providers of news content found little interest in journalistic news content, preferring instead such material as sports scores, movie reviews, and games. An executive participating in the project from the *Virginian-Pilot* and the *Ledger-Star* in Norfolk, Virginia, said users were "not looking for a newspaper on the system."[119]

Even so, the CompuServe project pioneered the way for dozens of other online newspaper projects in the early 1980s, creating a sense that urgent and important activity was underway. Said one industry consultant in 1984: "We are seeing a technological watershed—a sweeping away of long-established traditions and the opening of enormous business opportunity."[120] It is important to understand, however, that no one model was accepted as the way to proceed. The CompuServe initiative proved that such efforts were technologically feasible, but it provided little in the way of guidance about how much a company should invest in its own project. In the following examples, divergence of opinion emerges in terms of capital requirements. StarText represented a low-end technical approach, while Viewtron and Gateway are examples of multimillion-dollar initiatives.

StarText

The StarText project captured the attention of many within the newspaper industry—largely because of its simplicity. As much of the industry fixated on the pilot test underway for Viewtron (which will be examined next), StarText appeared as an alternative approach. The *Star-Telegram* in Fort Worth, Texas, partnered with Tandy Corporation to launch the online bulletin board system on May 3, 1982.[121] Although StarText insiders referred to their platform as video-text, it differed significantly from systems that touted graphics. StarText presented its content in a simple text format found in presentations known as Bulletin Board Systems (BBS). Unlike other bulletin board systems emerging at that time, StarText decided to forgo the commonly used pay-by-the-minute plan, opting instead to charge $5 per month for unlimited use. The service launched with a marketing campaign that promised immediacy and interactivity. The campaign featured the slogan, "The news you want when you want it."[122]

Nevertheless, six months after its debut, StarText had signed up only 50 users, largely due to major technical shortcomings. Its partnership with Tandy meant that users of other popular computers at the time such as Commodore, Apple, and Atari could not access the system.[123] And, as one of its early architects explained, StarText "wasn't truly an online system. You called the host computer and entered up to four requests; the host computer delivered the information, then hung up. If you wanted more, you had to call back."[124]

The *Star-Telegram* opted to keep the StarText project alive even with so few users, investing in the platform as a research and development project. In 1983, it ended the Tandy partnership and opened the service up to all comers. StarText continued to experiment with a variety of content offerings and improved its technology. By March 1986, the service had 2,200 subscribers willing to pay a flat rate of $9.95 per month, which covered the expenses for a staff of seven and allowed the service to break even financially. The capital invested in StarText through March 1986 had totaled $210,000, while operating losses before reaching the break-even point were $129,000.[125]

These financial numbers were insignificant when compared to the millions of dollars invested in the much higher-profile projects such as Knight-Ridder's

Viewtron. As a result, StarText inspired many in the industry to pursue a low-cost approach as a way to mitigate risk. The "Electronic Editions" of the *Spokesman-Review* in Spokane, Washington, is another example of this low-cost approach. Rather than install any new computer equipment, the newspaper transferred data to an existing back-up computer system and allowed users to dial in directly to this system. By using existing resources that did not require re-entering data, the newspaper said the system operated "for virtually nothing."[126]

By 1992, the StarText platform itself was licensed to other newspapers interested in deploying a low-cost bulletin board system.[127] But the market was about to change rapidly—events that will be chronicled in later chapters. StarText succumbed to the market forces of the Internet in March 1997—by then rechristened StarText Classic.[128] At its end, one of its editors eulogized the service, describing why StarText can be viewed as an important milestone in the development of online media:

> Online systems come and go, but this one was special . . . there weren't any fancy graphics, colors, or sounds; just plain old text and information and a "warmth" that will never be duplicated again anywhere. We were a small community of three or four thousand but we all knew what was going on with everyone. We were neighbors and we were concerned about each other. . . . We couldn't rely on fancy graphics or colors, we had to have content. Content was king![129]

StarText survived to reach the cusp of the Internet era not only because its operating costs were low, but also because the expectations for it by its owners were never couched in grandiose terms. The executives and editors responsible for StarText never seemed to invest in it the hopes and dreams of an entire industry. They saw it as an experiment, an interesting way to engage in an ancillary market during a time of long-term uncertainty. From that perspective, StarText is rarely viewed as a failure. The same is not true of the next two examples presented. Viewtron and Gateway were the subjects of significant press coverage and were allowed by their owners to become the standard-bearers for the newspaper industry's future.

Viewtron

When considered either by the notoriety achieved during its time or through the significance afforded it in historical retrospectives, Knight-Ridder's Viewtron system is an iconic element of the early 1980s media. As Boczkowski wrote: "no other development in this period illustrates the extent and character of videotex initiatives better than" Viewtron.[130] Knight-Ridder's management, convinced that the newspaper industry had to take a leadership role in determining the viability of electronic information delivery to consumers, launched a field test of the videotext system in July 1980, only a few days after the *Dispatch/CompuServe* test began.[131]

The company had begun developing its version of the British Prestel system a year earlier in partnership with AT&T, then the monopoly provider of telephone service in the United States.[132] In November 1983, Viewtron officially launched following an investment of $26 million and 14 months of field trials involving 204 homes in Coral Cables, a Miami suburb.[133] The significance the newspaper industry attached to Viewtron's launch was underscored when a spokeswoman for the industry's major trade association said: "There isn't anybody who isn't looking at it. In the minds of many people, if this doesn't work, there may not be a market for these kinds of services."[134] Viewtron's managers understood the level of scrutiny was extremely high. As one of Viewtron's vice presidents described it: "The whole world is watching South Florida. We are dancing naked on the stage of history."[135]

Knight-Ridder had not arrived at that point haphazardly. It had invested considerable resources testing the service before deciding to move forward with the project. Sigel reported that the households involved in the field testing leading up to the commercial launch represented 691 individual system users, a pool that provided enough positive data to prompt Knight-Ridder to move forward with the project.[136] He noted that Knight-Ridder kept much of its early market research proprietary, but was encouraged by several factors:

> usage . . . was 30 to 60 minutes per day per household;
> users weren't intimidated by the hardware;
> electronic messaging was a "key strength" of the system; and
> two-way shopping and banking had very strong appeal.[137]

At its launch, the graphic presentation was considered state-of-the-art for its time, and management appeared intent on selling the interactive functionality of the platform that had tested so well during the field trial. Features such as online banking, gaming, and shopping were heavily promoted. Twelve banks and more than 100 merchants participated in the initial roll-out.[138] There was sophisticated computer integration with several retailers. For example, consider this description from *Viewtron Magazine*, a publication sent to the subscribers of the company's *Miami Herald* newspaper to promote the new service:

> J.C. Penney replies quickly to orders from its catalogue because Viewtron connects you directly with Penney's order-entry computer in Atlanta. If the item you want is not in stock, the computer suggests alternatives. Perhaps it is available in other colors—you'll get a list of them.[139]

The system allowed customers to order merchandise by entering a credit card number on the screen and select either home delivery or in-store pick-up. With shopping services as well as other interactive features, including message capability, calendars, and bulletin boards, the Viewtron system reflected the application classes presented by Alber that were noted earlier in the chapter, and it offered—in concept—much of what contemporary consumers have come to expect from the Internet.

Even if the services offered by Viewtron were prescient of the Internet, the execution in delivering those services was not. Relying on a rigid and proprietary infrastructure, Viewtron failed to connect with its audience. Despite all of the hype and high expectations, Knight-Ridder shut down Viewtron in 1986 after attracting only 15,000 customers and losing $50 million.[140]

A major issue contributing to Viewtron's demise had nothing to do with content, but rather the system on which Viewtron was deployed. The system was developed in partnership with AT&T and initially relied on this company's proprietary videotext terminals for access. At the dawn of the personal computer era, a system designed for a proprietary terminal had limited consumer appeal, especially one priced at $900 when the system launched.[141] Ghosh noted that these terminals were later reduced in price to $600, but added that a major lesson from the Viewtron experiment and others like it "was that home consumers were unwilling to invest a large amount of money for special videotex terminals, that had limited functionality and power."[142]

When Viewtron attempted to expand the service outside of Florida, the AT&T terminals became a major point of contention with potential affiliates. An executive overseeing the project for the *Boston Globe*'s parent, Affiliated Publications, expressed the issue bluntly in the industry trade press, stating that the terminal cost was "far too high," adding, "There is no way you can have a home videotex system in which the cost of the terminal is $600."[143]

To offset consumer objections to the high terminal cost, Viewtron began offering a rental program that bundled equipment with 10 hours of service for $39.95 per month. The new pricing model worked for a time, resulting in "hundreds" of new subscribers.[144] However, consumers most likely to be interested in Viewtron were also those most likely to purchase a personal computer, and they saw no reason to also purchase or rent the proprietary AT&T terminal required for Viewtron access. Viewtron eventually offered an access kit for the emerging personal computer market, but the efforts were too late to save the system, Ghosh wrote.[145]

James K. Batten, Knight-Ridder president, tried to present Viewtron in a positive light even as he was announcing its closure. He claimed there had been "steady growth in the number of subscribers," but added that the use of the system among its subscribers had "not kept pace." He attempted to position Viewtron's closing as one of market timing rather than an indictment of the service's content offerings or the delivery execution. "Over time we might have been able to turn Viewtron into a viable business, but in weighing the continuing cost of investment against other competing uses for corporate funds, we decided it was in the best interest of shareholders not to continue," he said.[146]

Although Viewtron received most of the attention regarding videotext projects of this era, Knight-Ridder was not alone in creating a high-profile service only to close it due to lack of audience. The following section presents another example, Gateway, which was launched by Times Mirror, the parent company of the *Los Angeles Times*.

Gateway

Times Mirror launched Gateway as a system similar to Viewtron in 1984, expecting to showcase the service as an advertiser-supported medium. But Gateway could not gather enough audience to attract advertisers willing to pay enough to support the system. Times Mirror decided to close the venture in 1986.[147]

Although officially launched in 1984, the project actually had begun in early 1982 with pilot tests of 300 households in two suburbs near Los Angeles (Orange County and Palos Verdes) chosen for their "relatively large number of upscale, computer-hip Californians."[148] After the project officially launched, however, Gateway was unable to re-create its test market success across a broader market area.

From the outset, Times Mirror was clear about its motivations; it saw Gateway as a means necessary to protect its local market franchise. As the Gateway project was getting underway, Wright, the president of the *Los Angeles Times*, explained that his company believed "we should establish our newspapers as the foremost information providers in our local markets. . . . We shouldn't let others use our data bases before we consider how we could use it to our own advantage in future services that may develop."[149] He concluded with a familiar refrain among industry executives at the time; that doing something was better than doing nothing:

We are on the threshold of electronic information delivery. On the other side lie great opportunities for someone, including newspapers if we choose, and great threats if we do the wrong things. In my opinion, one of the wrong things would be to do nothing.[150]

But this executive was unique in that he asserted that newspaper companies may be unprepared for the work that was about to confront them, foreshadowing the problems that would bedevil Gateway and others.

He warned that applying the relatively new concepts of electronic publishing to online media would not go far enough. "Except for the application of some new technology in very recent times, we are doing the same thing we've been doing for decades. . . . We've used computer technology and offset printing to make the operation more efficient, but that is a change in how we do it, not a change in what we do."[151] This executive became one of the early proponents of creating new companies or new divisions within a newspaper organization to take on the responsibilities of the new online functions:

Many of us could fall into the trap of trying to manage our entry into the electronic field with an effort that is underfunded, understaffed and managed by people without experience as entrepreneurs. There's a lot to be said for establishing a separate group of people, adequately funded, with the single objective of developing a new line of business. . . . A separate organization could deal more effectively with potential partners on a basis that involves the newspaper only to the extent that it makes sense to the new enterprise, without forcing things into a shape that makes sense from a strict newspaper perspective.[152]

Times Mirror largely followed this model by investing relatively huge sums in the test project and forming a separate company with a dedicated staff of 60 employees at its launch.[153] Nevertheless, Gateway suffered from some of the same early technical shortcomings as Viewtron. The early Gateway project relied on customers using a television set to display content. However, customers complained about an awkward keyboard arrangement and difficulty of connecting with the service using the telephone.

The emergence of the personal computer led Times Mirror to abandon Gateway's television interface in favor of a computer-based application by mid-1985. But numerous other customer issues with the system plagued the project. Complaints included: complex billing that resulted in some customers incurring charges much higher than what they expected, slow connection speeds that made it difficult to render the system's complex graphics, system outages and connections that tied up the subscribers' telephone line while in use. Such problems contributed to "an enormous monthly turnover rate of roughly 60 percent [of subscribers]."[154] When Times Mirror closed Gateway, the service had about 2,000 subscribers,[155] but by then the project also had cost $50 million.[156]

The service's managing editor lamented the customer service issues as contributing to Gateway's failure, but he added:

> Our service was about as interactive as a Renaissance fresco. It's not enough to put canned restaurant and movie reviews, encyclopedias and recipes into a database, add some news and call it interactive. Subscribers told us they wanted to have personal contact with experts.[157]

He also said the Gateway experience held lessons regarding the size and type of audience necessary for conversation and community building:

> The "critical mass" necessary to create an electronic village never materialized because the decoder boxes were difficult to use. Then, when we discarded the boxes and appealed exclusively to computer owners, we ended up with a subscriber base heavily populated by techies, not exactly a village that noncomputer types wanted to visit."[158]

As Gateway and Viewtron closed in 1986, the newspaper industry was left to explain its actions. Industry insiders as well as analysts and investors wanted to understand why collective investments of well over $100 million had failed to produce any meaningful audience. Even the publisher of the *Star-Telegram*—the newspaper that created StarText—acknowledged that the failure of those two high-profile ventures had cast a "pallor" over the newspaper industry's videotext projects, but argued that his own experience with StarText should encourage the marketplace to understand that "videotex has a future."[159] But few in the industry took solace in the low-budget approach even as they pursued it. An industry that had wanted to be seen as leading the Information Society revolution was now viewed by many as cowering in retreat.

The Videotext Aftermath

In the early to mid-1980s, failures of videotext projects throughout the newspaper industry prompted executives who were expecting much larger audiences to question the viability of an online market. Executives accustomed to newspapers that attracted hundreds of thousands of subscribers could not fathom a business that required investments in the millions for a large-scale system, but produced consumer audiences in the low hundreds or thousands.

Knight-Ridder's Viewtron and Times Mirror's Gateway were the high-profile failures that prompted industrywide angst, but there were other examples that illustrate how quickly the industry launched these projects only to abandon them. KEYCOM, for example, was formed in 1982 as a joint venture of Chicago's Field Enterprises (which then owned the *Chicago Sun-Times* and WFLD-TV), Honeywell, and Centel, an independent telephone company.[160] KEYCOM, which was majority owned by Centel, spent nearly three years developing and testing a system at a cost of at least $11 million. But the project closed in 1985, only six months after its public launch. At the time of its closing, the service—called KEYFAX—had attracted only 100 paying subscribers.[161]

In another example, A.H. Belo Corporation, publisher of the *Dallas Morning News* in Texas, closed its videotext initiative after spending around $2 million on the project. The service had attracted only 200 subscribers. Belo said it closed the system because there were "just not enough terminals out there" to justify the continued operation, adding "when the terminals are there, we'll just come back. . . ."[162] An executive at the *Virginian-Pilot* and the *Ledger-Star* in Norfolk, Virginia, cited the same reasons when his company pulled out of the CompuServe project—too few computers in the market to justify the expense.[163]

In the aftermath of the failed initiatives, Henke and Donahue described videotext as an "interim technology."[164] They wrote:

> Consumers are hesitant to invest upwards of $300 for a technology which is not truly interactive and whose future is unclear. Conversely, videotex originators are reluctant to underwrite the cost of initiating a service and maintaining viable offerings while there is no industry standard and consumers appear to be reluctant to purchase decoders. Thus, false starts, failures and constant innovation lead to uncertainty in the marketplace which translates into financial inaction for all parties.[165]

Newspaper executives had ignored the lack of computing infrastructure in the U.S. during the rollout of these early experiments, and they overestimated how much consumers would be willing to pay at that time to acquire the necessary infrastructure.

An industry consultant said the newspaper industry based its videotext initiatives on a false assumption:

> There was a grand vision created, and the implicit assumption was that everyone wanted it. They spent a lot of money creating services, only to find the market didn't exist, and the thought of sticking around to develop them wasn't appealing.[166]

An article in the *New York Times* analyzing the demise of the three most prominent videotext services—Viewtron, Gateway, and KEYFAX—asserted that they "tried to offer too much to too many people who were not overly interested."[167] Reid Ashe, chairman of Knight-Ridder's Viewdata Corp. and the executive who managed the Viewtron project, concluded that "the market is thin and probably limited to the computer hobbyist," adding, "There's no prospect for it being a mass medium in the foreseeable future."[168] The newspaper industry based many of its assumptions on the overall consumer reaction to videotext services without analyzing the underlying dynamics of an emerging marketplace.

Raymond, however, discussed the failure of early videotext projects within a context of "mass market phenomena," which he said involves "two phases: the *aficionado* phase and the *trend* phase."[169] He explained:

> The *aficionado* phase revolves around a group of people who share a common interest, activity or behaviour, which is facilitated by the use of a good or family of goods. . . . Aficionados are often individualistic, opinionated, and single-minded in devotion to their interest. . . . Aficionados often congregate in groups, and in doing so develop notions of status. The common interests of the group implicitly define goals or activities which are generally approved by the aficionados, and which distinguish them from non-aficionados. Those who adhere most closely to the group's goals and methods, and especially those who show uncommon creativity or endurance in doing so, are accorded special status within the group. Status is also conferred on specific goods or services which are deemed to incorporate or exemplify those characteristics which the group extols. The *trend* phase . . . begins when the status originally derived from the goals and activities of the aficionados becomes a characteristic of the goods and services they prefer. Status can then be obtained by merely acquiring the goods. In the trend phase, large numbers of non-aficionados attempt to obtain status by acquiring those goods the aficionados deem appropriate. . . .[170]

Raymond theorized that, based on these inner workings of human nature, developers of the early videotext systems not only failed to attract aficionados, but were "antagonistic" to them.[171] He concluded:

> At every turn videotex developers made choices which would discourage aficionados. . . . Rather than enhancing their customers' prowess by providing at least the appearance of power and complexity, videotex is boringly simple. Rather than provide variation and attendant status, videotex is the same vanilla flavour everywhere. Rather than permit rapid change and development, videotex concentrates

on "standardization." Rather than permit individualism and distributed control, videotex centralizes. In effect, videotex is an attempt to start a trend without aficionados.[172]

While Raymond contended that videotext was too simple to attract enough early adopters, others found the technology too complicated to interest average consumers. A usability study of the Viewtron system conducted in 1984 concluded that "use of Viewtron proved to be fairly complex."[173] The researchers wrote of their findings: "Subjects with more computer experience had more positive experience with Viewtron and those indicating apprehension about computers in society on the pretest found Viewtron to be more confusing, irritating and complex."[174]

When newspaper companies began their quest for electronic distribution in the early 1980s, most of the comments regarding videotext as the underlying technology were positive. When issues began to surface regarding the lack of audience traction as early as 1984, critics began to question the videotext technology itself. A comment from an analyst with the Yankee Group is representative of such sentiment: "The problem is that its technology is 10 years old. That first generation of videotext has the stench of death about it. It really does give boredom a new meaning."[175]

In this way, the early videotext efforts clearly suffered from a market timing problem. The earliest forms of videotext were designed for proprietary hardware in part because the personal computer had yet to gain a significant foothold in the general consumer market. But when computers started to gain traction, videotext services were slow to adapt. And when they did convert, the designs aimed at a mainstream audience did not hold enough fascination for the early computer adopters.

The issue of market timing was a central theme addressed in a Viewtron retrospective produced by the Poynter Institute. On the 20th anniversary of Viewtron's commercial launch, this newspaper industry research and educational organization convened an online roundtable discussion with several former members of the Viewtron staff in an attempt to put the project in historical perspective. The organizer of the event wrote of Viewtron's timing:

> It was the right idea; wrong decade. . . . For all its faults, there were important lessons learned from Viewtron. It was a bold attempt to change the rules of getting information to the consumer. It provided a glimmer as to the changing nature of how consumers get and use information. . . . Ultimately, Viewtron was probably the wrong technology at the wrong time.[176]

However, if Viewtron represented only a lesson in deploying the wrong technology at the wrong time, it likely would not have received the attention it has. The importance of Viewtron, as the Poynter event organizer noted, stems from "the conversation it started in the newspaper industry about the future of the communications business."[177] From this perspective, the edited transcript of the roundtable discussion provides interesting insight directly from several key staff members involved in the management of Viewtron and its content offerings.

For example, Phil Meyer, a journalism professor at the time of the roundtable, earlier had served as Viewtron's director of market research. He addressed the shortcoming of Viewtron's centralized nature:

> We made the mistake of thinking in newspaper analogies. Thus the central computer was like a printing press in our minds, and telephone wires were the delivery trucks. We never foresaw anything as free and open as the Internet or grasped that there would be no central computer. As newspaper people, we were looking for a community-based natural monopoly, like a newspaper, but without the variable costs of paper, ink, and transportation.[178]

This observation echoes the theme from Raymond, who had postulated that centralized control had contributed to the failure of videotext because it emphasized standards to the point of making the system unappealing to individuals who wanted some element of control.[179] In the case of newspaper executives, however, that element of central control and standardization was a major reason videotext appealed to them as a technology.

The attempt to operate Viewtron with the central control model of a newspaper contributed to its problems, and represented a "failure to think farther outside the newspaper, local-monopoly box."[180] The retrospective discussion revealed that Viewtron subscribers spent 80 percent of their time online with interactive features such as email, bulletin boards, and study guides. Because a large portion of the services revenues were derived from time-based online charges, this posed a business model dilemma. As one former editor stated, "Viewtron was spending about 80 percent of its budget to create news, which generated less than 20 percent of the revenue. That's when [Knight-Ridder] executives decided that this was not a news medium and they wanted to continue to be a news company."[181]

Ashe, Viewtron's top executive, agreed that such assessments contributed to the eventual decision to close the operation. He said that usage had not conformed to company expectations, stating that "Viewtron died from an excess of funding and expectation. Too much was invested in too specific a vision, and there was no appetite to change course. . . . The more closely we approached a viable service, the less it looked like a newspaper."[182]

Technical shortcomings of videotext as it was deployed and slow network connections available in the early 1980s contributed to Viewtron's failure. However, the comments from key former employees participating in the retrospective discussion help explain its demise in ways that are less generally understood. Viewtron's creators saw the potential of an array of consumer options, including auctions, educational courses, travel services, banking, online shopping, and games. These services were heavily promoted in Viewtron's marketing. Therefore, one can see that, intellectually, Viewtron was built to showcase the capabilities of an interactive network. Emotionally, however, the company behind Viewtron was unable to shed the news bias associated with its newspaper roots.

The comments of those involved reveal that resources to support interactive functions never reached the levels of the resources allocated to support news. Viewtron's news and information components received the largest share of budget even though consumer usage patterns did not warrant it. The concept may have represented a paradigm shift, but the company attempting to exploit it could not adjust to a new way of organizing resources to take advantage of a new medium. Viewtron's former managing editor said, "we all had notions that Viewtron could be an extension of the newspapers."[183]

When that failed to materialize in the way the company wanted, the decision was made to close Viewtron rather than recast the resources and build a new business around the interactive services. Ashe commented, "We should have started over, but that wasn't an idea that we could sell."[184] The actions of Knight-Ridder, Times Mirror, and other newspaper companies at this time revealed an industry not as prepared to change as many observers thought at the beginning of the decade. Boczkowski, in studying the end of Viewtron and similar projects, concluded that newspaper companies lacked the "conviction" needed to embrace a new medium that would force changes to existing business models and processes. He wrote:

> . . . in a context marked by information society rhetoric, the ideology of technological determinism, and the preexisting computerization of the industry, newspapers appropriated videotex from the standpoint of a publishing mindset that had historically evolved over a couple of centuries of producing content for a large number of readers. Thus, they pursued videotex less out of a conviction that they needed to alter their production procedures and values to create an entirely different media artifact than because this was something they "had to do."[185]

Knight-Ridder launched Viewtron in part because it wanted to maintain what it considered a leadership role in the information industry at the dawn of an era that would be defined by the emergence of electronic publishing. By closing Viewtron in the manner in which it did, Knight-Ridder was content to have its newspaper business remain closely associated with print rather that the broader information marketplace. Kyrish recognized the paradox represented by the newspaper industry's foray into electronic distribution through videotext. She wrote:

> A repeated notion in the early and mid-1980s was that companies needed to keep pouring money into videotex development, to be "ready" when the market hit. Instead, a paradox occurred: early videotex spending was based in part on the expectation that interactive services would threaten the profits of traditional media companies. So companies created videotex in order not to be caught unprepared by videotex.[186]

Viewtron, Gateway, and similar projects of the early 1980s were recognized as media vastly different from the newspaper companies that spawned them. Their failures stemmed from a disconnection between the intellectual construction of these projects as interactive services and the emotional desire for them to become extensions of a newspaper.

Discussion

Examining the newspaper industry's reaction to the videotext era is important because of the influence this period had on future decision making. The newspaper industry's attempts to reconcile internal conflict regarding the nature of online newspapers will be examined further in subsequent chapters that explore the industry's relationships with proprietary online services and the emergence of the Internet. Many of the issues that surfaced with the early videotext projects and discussed in this chapter—such as the degree of interactivity offered and the potential transition of classified advertising into a database-driven model—remained prevalent as newspapers confronted new technology platforms in the 1990s and 2000s.

Initially, many newspaper industry leaders chose to view the closing of the early videotext projects not as an industry failure, but as a positive referendum on the future of printed newspapers. Boczkowski noted, for example, that the "commercial failure" of the high-profile videotext projects led many in the newspaper industry to become "reassured . . . about the viability of print."[187] Later, the financial losses incurred by these projects were recalled as excessive and contributed to the industry's reluctance to take risks when confronting decisions regarding new technology. Both attitudes guided the industry toward a protectionist agenda, and it spent considerable resources in the 1980s and early 1990s lobbying the government and regulators to keep potential competitors out of the online information business in local markets. This aspect of the newspaper industry's relationship with online media will be discussed in Chapter 3.

Before that exploration, however, Chapter 2 examines the newspaper industry's relationship with cable television in the early 1980s. Although the newspaper industry was engaged in videotext exploration, it was also enamored with the distribution possibilities inherent in cable television. Newspaper companies were early proponents of the cable industry's transformation from a rural delivery system to a national platform for programming diversity. An examination of the newspaper industry's involvement in cable television in the early 1980s is important because it corroborates the experience with videotext. Newspaper companies were eager to invest in the promise of Information Society rhetoric at the outset, but were quick to abandon those investments when the market did not develop as they anticipated.

2

The Newspaper Industry's Brief Cable Television Strategy

Leaders of the U.S. newspaper industry in the early 1980s were convinced that emerging technologies would alter the delivery of information, and in turn, change the nature of the marketplace in which their companies operated. But while there was consensus that new technology would bring significant change, a wide range of opinions existed about the form such technology would take and how the newspaper industry should respond.

The previous chapter focused on the newspaper industry's videotext projects in the early 1980s and explored how those experiments represented important early attempts by the industry to exploit interactive media. This chapter also examines activity from the early 1980s, but focuses on the newspaper industry's flirtation with cable television. A cable television trade association official described newspaper executives at this time as in "a panic about cable television."[1] Therefore, it is important to examine the influence a burgeoning cable television industry had on the decisions made by newspaper companies regarding technology and electronic distribution in the early 1980s.

Collectively, newspaper companies spent more of their capital resources on cable television ventures than on any other form of emerging technology during this period. This chapter explores how the early development of cable television resonated with newspaper publishers who adopted the notion that a wire into every home could be a way to protect their interests in local markets. This chapter shows how the newspaper industry initially gravitated to cable television for several reasons: it had a more proven business model than videotext, early cable entrepreneurs were eager partners seeking access to newspaper content, and cable television companies were seen as a less threatening alternative to the established telecommunications monopoly.

This chapter includes a brief review of cable television's history including some relevant statistics that provide market context for the decisions newspaper publishers made in the early 1980s. The chapter includes examples of specific newspaper company involvement in cable television to illustrate the variety of projects undertaken and the range of investment involved. However, when these investments did not deliver the expected results, newspaper companies made a quick exit from their cable projects. This chapter discusses some of the reasons for the market withdrawal and critically assesses the ramifications of those decisions. The newspaper industry's direct involvement with cable television was relatively short, but the influence of this activity was substantial. The withdrawal from cable television combined with the high-profile videotext failures discussed in Chapter 1 resulted in a newspaper industry that was more insular—a trait that contributed to a protectionist stance and a political war with the telecommunications industry that is the subject of Chapter 3.

Cable Versus Computers

There was much discussion in the nation's popular press in the early 1980s about the coming "wired" society. But exactly how the nation would become wired was still open to considerable debate. The ubiquity of the personal computer that we take for granted today was still science fiction at that time. In 1978, only about 150,000 personal computers had been sold for use in the home, primarily to electronics hobbyists.[2] The Computer Industry Almanac reported that in 1980 there were only 10 personal computers in use for every 1,000 people. IBM contributed to a sales surge when it launched its personal computer in 1981, but the devices' penetration into society remained relatively small throughout the early 1980s. According to the Computer Industry Almanac, 3.8 million personal computers based on an IBM standard were sold in the U.S. between 1981 and 1985.[3] When also factoring in Apple sales, the per capita statistic climbed to 99 personal computers for every 1,000 people in 1985, but still represented slightly less than a 10 percent penetration of the U.S. market.[4]

The newspaper industry tried to assess how the adoption of personal computers would affect local markets. A June 1984 study by the director of information technology at the Associated Press received considerable attention within the industry because it seemed to counteract the excited rhetoric surrounding the personal computer with a dose of statistical reality. The study estimated that, on average, only 1,900 personal computers were deployed per individual newspaper market.[5] This figure led the researcher to remark, "There are probably papers who have more newsstands than that."[6] In reacting to the study, the newspaper industry's leading trade association said that it understood personal computer deployment was mostly a business phenomenon, adding that the personal computer represented little opportunity for general-circulation newspapers.[7] The videotext projects discussed in Chapter 1 reflected this sentiment as most initially deployed on proprietary terminals rather than personal computers.

Therefore, in reflecting on this era in a critical analysis, it is important to understand the limited availability of personal computers in the consumer market. As Boczkowski observed:

> Although personal computers became the dominant alternative by the mid 1990s for reasons that seem quite logical from today's standpoint, this knowledge should not be used to read history backwards: none of these delivery vehicles seemed an obvious choice for the actors struggling to make sense of an utterly complex and uncertain situation.[8]

Given the historical context of personal computers, it is much easier to understand how the newspaper industry in the early 1980s was more enamored with cable television as the platform destined to usher in the information age. Newspaper publishers had a familiarity with cable television that did not exist with computers.

Although the notion of a wire to every home may have been a subject of the revolutionary rhetoric, cable television itself had been around for decades. Newspaper publishers were comfortable with cable television as an established industry seeking to exploit market opportunity just as they were. The next section of this chapter recounts a brief history of cable television, which is important background for understanding how the business relationships with newspapers emerged.

A Brief History of Cable Television

Cable television originated as a mechanism to deliver television signals into areas that could not receive a clear picture through broadcast antennas. The business of delivering television over a wire began in the late 1940s, but researchers have disagreed as to where the first system launched. Some researchers attributed the birthplace of cable television to Astoria, Oregon, in 1949, but others credited Lansford, Pennsylvania, in 1950 or Mahoney City, Pennsylvania, in 1948 as having the first cable television system.[9] In any case, cable television had evolved into a respectable business over approximately 30 years, generating revenue of more than $1.5 billion by the late 1970s.[10]

U.S. government statistics showed the industry grew significantly during the 1970s, increasing from 2,639 systems serving 5.3 million subscribers in 1971 to 4,225 systems serving 15.5 million subscribers in 1980.[11] That number of subscribers represented 20 percent of the nation's households at that time, and the industry was aggressively trying to sign up another 20 percent of the nation's households that were not subscribers even though service was available to them.[12]

Although the 1970s had seen significant growth for cable television, subscribers were mostly in rural areas because city dwellers lived close enough to broadcast transmission towers. They did not need a service to receive television that could be obtained through a quality antenna. Cable television operators realized, however, that future expansion required finding ways to attract interested urban and suburban customers that went far beyond the appeal of signal quality. The expansion of satellite technology provided the industry with the infrastructure necessary to increase its programming and mount a strong push into metropolitan areas.[13] Moss described the status of the cable television industry at this time:

> The emergence of new technologies, linking computer systems and communication satellites with the home, has made cable television more than a simple mechanism for improving the reception of broadcast television systems. Cable television can not only expand the number and type of television programs; it can also allow two-way or interactive communication to occur. The interactive potential of cable has fostered much speculation about the public and private services that two-way systems could provide.[14]

This notion that cable television systems could transform the television viewing experience through two-way communications was largely responsible for changing the way regulators viewed this new platform's role in the media market.

Although the growth experienced by cable television in the 1970s was substantial, it could have been even greater. For most of the 1950s, federal regulators had taken a hands-off approach to the cable television industry because they felt it posed no competitive threat to the broadcasting television industry. By the mid-1960s, however, regulators saw cable television as having the potential to "seriously threaten the viability of the over-the-air television system."[15] As a result, cable television's growth had been intentionally restrained in the 1970s through regulatory efforts of the Federal Communications Commission (FCC). By the 1980s, however, the FCC had reversed course and put in motion significant regulatory reforms that propelled cable television into significant new markets. The regulators had determined that it was no longer in the best interest of consumers to restrain a technology that promised programming diversity and interactive services. As the 1980s opened, the National Cable Television Association (NCTA) proclaimed: "We are a booming industry."[16]

Cable Television and Interactivity

The cable television industry had been savvy enough to recognize that the climate for regulatory reform had been enhanced by the surge in Information Society rhetoric. Positioning its technology platform as one that could deliver on the promises of the rhetoric through interactive capability allowed the cable television industry to differentiate itself from traditional broadcast television. By establishing its technology as much more than a platform for retransmission, cable television elevated its standing beyond that of a traditional medium.

Besen and Crandall agreed that much of the regulatory relief was due to the technological advancements cable television represented. They described cable television as "an excellent example of how difficult it is to restrict entry when technology is changing rapidly." They acknowledged that some regulation was inevitable, but argued that the compelling nature of the technology available through cable television "made it extremely difficult for the [FCC] to continue to constrain cable growth."[17] Besen and Crandall concluded:

> This has occurred because new programming services, distributed by a new technology, are replacing imported broadcast signals as the most attractive offering of cable television in the larger markets. Cable television is no longer the enhancement of local broadcast signals. . . . It is now a service which offers a much wider array of services—an array which will increase during the 1980s.[18]

The promise of two-way capability contributed to the emerging perception that cable television had "inherent advantages" in the media marketplace.[19] The ability of viewers to communicate upstream with a cable system—to respond to programming choices and to make the decision to purchase programming on a pay-per-view basis—was viewed as a powerful innovation. Noam wrote:

In its commercial potential, two-way communications is a marketer's dream come true, since consumers can respond to advertising messages instantaneously by pushing buttons to make an order and to transfer funds in payment. . . . Cable's two-way capability also makes possible services which should be as useful to consumers as they are profitable to business enterprises: alarm systems, meter reading, electronic banking, video text information, classified ads, and many more. Consumers will therefore benefit from two-way cable as a communications medium quite apart from its entertainment content. . . .[20]

With the potential for such a smorgasbord of services, investors saw multiple revenue streams that would allow them to generate the cash required to offset the capital investments needed to build cable television's infrastructure.

Companies quickly established systems and claimed specific geographic territories. A spokesman for the NCTA said, "People are running around looking for franchises and to get the cable systems they hold franchises for into operation."[21] The investment activity reached such fevered levels that some described it as a 1980s-style "gold rush."[22]

Newspaper Companies and Cable Television

Many newspaper companies were attracted to the cable "gold rush" for purely financial reasons. Cable television was viewed as a way for a business to make a sizeable return on a reasonable investment of capital. But some newspaper companies were attracted to cable television for strategic reasons as well. They were impressed by a business model that had enticed millions of consumers to convert from free television to a subscription model. Radio and television was an advertiser-supported media, but the notion that consumers were willing to pay extra for specific television programming changed the way traditional media thought about electronic media delivery. An editor explained, "Cable enhances the amount of information, at a cost small enough to create a massive market."[23] Newspaper companies wanted to be part of such an ecosystem and were lured by the expectations of ancillary product offerings they could bring to the party.

Text-based news services and classified advertising channels initially were thought to hold the most promise. Newspaper publishers believed that their existing news and sports content, along with their lucrative classified advertising business, could be redeployed to create text-based services. Cable operators thought these types of channels would be attractive to consumers interested in a source of local market information. In these early years, making this type of content available was not an issue, given that systems had more programming needs than their limited video programming could fill.

Initially, cable television companies encouraged newspaper companies to become partners. Time Inc.'s cable division, for example, sought out partnerships with newspapers "because of their established position in the community."[24] Time Inc. acknowledged that newspaper publishers were confronting decisions about where to allocate resources in the early 1980s as they contemplated electronic distribution and presented cable television as their best alternative. A Time Inc. vice president stated that "cable television is better positioned for text services than either the broadcaster or the telephone company." The executive appealed to newspaper companies to become partners because he said they would bring "ideal journalistic and production skills for this medium."[25]

Newspaper publishers responded favorably to such deference as they turned to deals with the cable television industry as a preferred method for entering the era of electronic distribution. As the following section explores, newspaper companies invested millions of dollars in cable television arrangements, but some critics argued that newspaper companies did not drive the hardest bargain as they were eager to expand into the exciting new arena.

Cable Television Deals

The newspaper industry had dabbled in cable television since its beginning, and by the end of the 1970s, newspaper companies "had some degree of . . . ownership" in slightly more than 12 percent of the cable systems then in operation.[26] However, the newspaper industry's involvement with cable television escalated in the early 1980s as the two industries sought ways to exploit the strengths of the other party. Cable television executives initially wanted access to newspaper content, while newspapers envisioned that a cable system's local market franchise could be exploited as an extension of their own local market monopolies.

From 1981 through 1984, U.S. newspaper companies responding to industry surveys reported that they invested considerably more capital in cable television ventures than in their initiatives with non-cable videotext. These U.S. newspapers said they invested $38.6 million to either purchase or lease cable television infrastructure from 1981 through 1984. In comparison, these newspapers said they invested slightly less than $13.5 million in non-cable electronic projects such as videotext during this same period.[27]

Cable television systems are highly dependent on capital investment to build the required infrastructure. This was especially true in the early 1980s when large portions of the country were yet to be wired. Trade association estimates during this period reported that it cost between $10,000 and $15,000 per mile to deploy cable television connectivity in typical areas. However, those costs could soar in large cities such as New York to as much as $150,000 per mile. The capital-intensive nature of the business tended to drive up prices for systems, which were typically selling in 1980 for between $500 and $700 per subscriber. The selling price of some systems spiked to as high as $1,000 per subscriber.[28]

Newspapers large and small were eager to participate in the burgeoning cable market, but those prices seemed steep to some publishers. That led many newspaper companies to lease channels on existing systems rather than purchase a direct ownership stake in a cable television company. In some cases, cable operators provided channel capacity to newspapers for free in exchange for content services.[29] The newspaper industry's trade association reported that newspaper deals with cable television ranged from outright ownership of systems to channel leases that most often involved the creation of a profit-sharing joint venture. However, the trade association's legal counsel worried that some newspaper companies eager to join the cable television frenzy were not structuring their deals in the best way. In some arrangements, the association's legal counsel maintained that newspaper companies failed to negotiate a fair value for the content, opting to perform "almost a wire service function for the cable operator in providing him with information that he in turn sells to his subscribers."[30]

Industry leaders worried that such arrangements would set the market value for newspaper content too low and diminish the influence newspapers could bring to this emerging market. The trade association advised its member companies to craft a strategic plan and proactively decide how they wanted to approach their cable television endeavors before engaging in negotiations with cable system operators.[31] As cable television embarked on a decade of remarkable expansion, its companies took the upper hand in its dealings with the newspaper industry. As one consultant described it, "Cable's grade in conducting business with newspapers has typically been an 'A'; newspapers rarely have deserved more than a 'C.'"[32]

Understanding how unbalanced relationships ensued from the negotiations between newspaper companies and cable television companies requires exploring more of the rationale for why newspaper companies entered into these deals. The following section more closely examines the many reasons newspaper company executives gave for entering the cable television market in the early 1980s.

Newspaper Industry Rationale

The publisher of a small daily newspaper in Iowa is an example of an industry executive who pursued cable television opportunities by leasing channels on local cable systems. He used the channels to provide text-based local news and advertising services, and his efforts were followed closely by other newspaper companies who hired him as a consultant. Gerald Moriarity, who was publisher of the *Globe-Gazette* in Mason City, Iowa, had a message to other newspaper companies that was about warding off competitors: "I warn them if they don't try to get involved, they sure as hell will look up one day and see that a competitor is in the driver's seat."[33] Most publishers seemed to believe that newspapers and cable television represented media that, if deployed properly, would complement each other rather than detract from either.

A study by the newspaper industry's principal trade organization found that 69 newspapers in the U.S. were involved with commercial information services over local cable television systems at the end of 1981, representing a potential market of more than 1 million subscribers.[34] This study concluded that "many other newspapers are exploring opportunities for similar ventures" because they were attracted to cable television over other forms of videotext applications. The reasons listed for favoring cable television included:

1. Relatively low cost of entry.
2. Higher likelihood that the services can operate at a profit.
3. Built-in subscriber base meant the services were not marketed as a stand-alone product.
4. No special equipment beyond the cable box was required by the consumer.
5. Services were included as part of the flat monthly cable fee (telephone-based textual information services at that time typically charged based on the time users spent with the service).[35]

Another study by the same organization in 1983 found that newspaper companies were reacting to competitive fears, but also acting out of desire to learn about new technology. When participants were "asked to choose which of two reasons for getting involved in a cable venture was more important to their newspaper, 34 newspapers, or 49 percent indicated it was a defensive, competitive move, compared with 25 that indicated that it was because cable seemed like a good business opportunity."[36] When the study also asked participants about other reasons for entering the cable television market, "42 newspapers, or 60 percent, indicated that it was to prepare for future communications technologies such as teletext and videotex. . . ."[37] The underlying notion of preparing for the future as expressed by these survey results played a significant role in how newspaper companies negotiated their agreements with cable television operators. However, the increasing sense that cable television would become a major media force also influenced the relationship newspapers had with cable television.

Becker, Dunwoody, and Rafaeli argued that the promise of cable television relied on "the notion of increased offerings of program content." But they added that "despite the simplicity of this observation" no one had any clear understanding of how more programming would disrupt audience patterns and the consumption of other media.[38] Nevertheless, they predicted that as cable television expanded, it would siphon away consumers and their financial resources from other media platforms. The researchers contended that media primarily compete on two fronts: "financial resources" and "time."[39] Given that cable television's subscription-based business model would capture financial resources and its additional programming would divert time, the researchers concluded that cable television represented a very significant competitive force. Working within the context of media usage models presented by others (McCombs,[40] McCombs and Eyal,[41] Weiss,[42] and Robinson[43]), Becker, Dunwoody, and Rafaeli stated:

> . . . cable ought to have an impact on other media use activities if for no other reason than that it provides content similar to what the other media are providing. In fact, cable technology has the capability to provide almost all of the content now being provided by the other media individually. The impact of cable ought to be the strongest. . . .[44]

Newspaper executives in the early 1980s clearly recognized the same factors as the scholarly researchers who were studying media markets: cable television represented a new paradigm for media content distribution. However, newspaper executives also acknowledged that they did not fully understand how they would exploit the new paradigm. Until that could be determined, newspapers were reluctant to enter into short-term agreements with cable companies. There was a genuine fear by publishers that cable companies would extract value in the short term only to shut them out over the long term. The result was a dogged pursuit of long-term, multi-decade agreements.

For example, the *Recorder* in Amsterdam, New York, in 1980 signed a 10-year lease with an option for another 10 years to lock up a channel on its community's cable system. The newspaper's publisher said: "The bottom line is that we have positioned ourselves for whatever is coming along regarding a tie-in between the newspaper and the electronic media. That's really what it's all about."[45] The publisher of the *Shawnee News-Star* in Oklahoma echoed the same sentiments after locking his newspaper into a 13-year channel agreement with the local cable system: "We still don't know where it is going. What we do know is that wherever it goes in Shawnee for the next 13 years, we're going to be the owners of it."[46] There is no way to know whether these specific deals were among those that caused the newspaper trade association's angst discussed earlier, but they are representative examples of how newspaper companies believed that long-term contracts were essential to protecting their investment in cable television operations.

Essentially, newspaper executives traded away financial incentives that could have made such arrangements more lucrative in the short term as they sought and won long-term agreements with cable television operators. The newspaper in Shawnee, Oklahoma, agreed to an escalating payment scale in which the cable operator received 10 cents per subscriber in the first five years, 15 cents during the second five years of the agreement, and 20 cents during the last three years. The cable operator agreed to accept half of the pay as promotional advertising in the newspaper, which allowed the newspaper to mitigate its actual cash outlay.[47] Newspaper deals with cable companies varied widely, and while this is an actual example, it is not possible to characterize it as a typical deal. In many cases, neither party would agree to absolute dollar terms as was the case in the Oklahoma market example. Instead, they would agree to share in revenue generated on a percentage basis.[48] Under revenue-sharing arrangements, cable operators believed they were better positioned to generate more revenue if newspaper-operated channels were successful and newspaper companies felt they were protected from being locked in to set payments should the channels fail.

Even though the newspaper industry may have lacked long-term clarity about its plans for cable television, the agreements that were signed in the early 1980s required immediate action. The following section focuses on several issues surrounding the execution of those agreements in the short-term, including a brief discussion about how the journalistic culture of newspapers influenced how these agreements unfolded.

Short-Term Execution

Reflecting the names of other electronic information technologies of this period—videotext and teletext—the textual display of information on cable television systems was called cabletext. The Time Inc. executive who spoke about the opportunities for newspapers in cable television warned potential newspaper recruits that information displayed on a television represented a different medium. He told them that cabletext required skills to edit information tightly "since the small screen is not an ideal reading medium."[49]

Newspaper companies that ventured into the medium early quickly learned that the challenges of producing cabletext material were not exaggerated. Consider the following description from a newspaper editor assigned to a Knight-Ridder Inc. cable television project at the company's *Lexington Herald-Leader* in Kentucky:

> In cabletext, we're talking about a maximum of eight 32-character lines that will be on the screen no more than 17 seconds. Yet, this eight-line story that will be on the home screen for just 17 seconds must first attract the interest of the viewer and hold it long enough for the viewer to read and comprehend the message.[50]

Such writing required skills were more closely aligned with advertising copywriting than news writing, and culture debates soon erupted between editorial managers and newspaper business executives. The editor at the *Lexington Herald-Leader* wrote that he was often asked of cabletext, "Is it newspapering?" adding that his answer was cabletext "can deliver only the 'who, what, where and when' of journalism, leaving the 'why' to the newspaper."[51] Actual content issues were only part of the reasons why newspaper editorial and business managers clashed over cable television ventures. The way newspaper publishers chose to staff these operations exacerbated the problem.

Many newspaper companies that launched cable projects also were involved with videotext projects. They were allocating resources between these ventures as a hedge against an uncertain future, but they also wanted to at least break even if not turn a profit in the short term. The results were operations with very few employees. Writers and editors who handled news were also expected to produce and manage advertising content. This breached the time-honored code of keeping advertising and editorial operations separate, referred to in the industry as the separation of church and state.

Nevertheless, the *Lexington Herald-Leader* editor acknowledged that his cabletext news staff often contacted banks, for example, to update interest rates in advertising displays. They also coordinated advertising content directly with sales personnel. "Such details would drive many a newspaper copy editor up a wall. In cabletext, they're just part of the job," he said.[52] Editors were concerned that the pressure for newspapers to find their way in with electronic information delivery had led publishers to forgo the industry's journalistic principles too easily. These concerns spilled over into how the very deals themselves could jeopardize a newspaper's journalistic integrity.

Some of the journalistic concerns about newspaper and cable television deals involved the franchise process. In the early days of cable, companies were awarded franchises by local governments often following intense, politically tinged negotiations. Cable operators worried that a newspaper's political coverage could affect its franchise status, while some newspaper editors were concerned that business pressure from a cable partner over such coverage could threaten the newsroom's independence. Moreover, many editors were concerned that providing content on cable systems that were operating under a franchise granted by a local government opened newspapers up for new government scrutiny and possible regulation. There were editors who felt that providing content on such platforms was tantamount to inviting the government to intrude on the newspaper industry's freedom.

Others, however, saw the increasing use of electronic distribution as inevitable and believed that the newspaper industry had to exercise leadership to see that First Amendment concerns were protected. Wicklein framed the journalism interest concerns:

> In this country, free flow of information is going to be determined by how the First Amendment is applied to the new technologies of communication. The First Amendment provides two things: the right to express ourselves freely, and, implicitly, the right to know. It encourages the widest diversity of ideas available to the listener or the reader. . . . The time is coming when both these rights will be exercised *primarily* through two-way, electronic communications.[53]

Wicklein challenged the newspaper industry to call for revisions to communications laws specifically to include First Amendment protections. He concluded that "If we are to guarantee free flow of information in the new technologies, one further step is essential: Federal agencies, especially the [Federal Communications Commission] have to be prevented from interfering with content."[54] Wicklein and others of a similar mindset wanted to ensure that cable television platforms were treated by regulators as common communications carriers. They wanted distribution technology to remain separate from the content it delivered similar to the way telephone systems were separated from the voice communication they transmitted.

Newspaper industry leaders in these early years were much more concerned about the separation of content and technology in relationship to the nation's telephone infrastructure than with cable television. Wicklein conceded that cable television operators had established precedent early through myriad content services, noting that they would unlikely "give up their right to control program and information services easily."[55] This issue as its concerns telecommunications will be addressed in Chapter 3. It is important to note here, however, that the newspaper industry's early views regarding how cable television should be regulated were tempered by its own companies entering into ownership deals and long-term leases with cable television companies.

It appeared that many newspaper companies wanted to have it both ways—guaranteed access for newspapers to the cable systems, but also the ability to enter into exclusive deals to keep other competitors out of the business. Wicklein called access for newspapers to cable television systems "essential" to the free flow of information, but he noted that "this does not mean they should be allowed to negotiate exclusive contracts."[56] For all of the noble First Amendment rhetoric that accompanied the early discussions of newspapers and electronic distribution, the actions of newspaper companies underscored that they were businesses protecting their own self-interest.

Cable television companies also were motivated by their own self-interest and were intently interested in market expansion. Even though cable companies initially had sought out investment deals and other business partnerships with newspapers, the cable television industry's interest in such partnerships changed significantly as technology advancements and new programming services allowed them to expand their television content without the help of newspaper companies. The following section explores how the rapidly evolving market affected the relationship between newspaper companies and cable television operators.

A Rapidly Evolving Market

Cable television operators began to replace their early systems that were limited to only 12 channels in many cases with more sophisticated technology capable of delivering dozens of channels (more than 100 in some advanced systems). As these new systems were unveiled, the nature of the market changed along with the underlying technology. Through the success of premium programming channels such as Home Box Office (HBO) and the favorable prospects of upstarts such as the CNN and the Weather Channel, cable television had been transformed from merely a distribution platform into a recognized purveyor of original media content.

As such, cable operators saw less need to give up valuable channels to text-based services. By 1983, there were more than 100 cabletext services underway in connection with newspapers,[57] but dozens of other newspapers wanted to enter the market. Those newspaper companies that wanted space on cable television systems in 1984 found a different negotiating climate given that cable television system owners were reluctant to lease a channel for textual purposes. They no longer wanted to enter deals that would encumber channel capacity at a time when the long-term prospects for video programming appeared to be much more lucrative.[58]

Moreover, cabletext services that had launched were having difficulty attracting an audience. Content scrolling on a television screen had failed to excite consumers who largely ignored those channels even when they were offered for free. Consumer reaction to the services made it abundantly clear that there was little market, if any, for those services to be offered as premium channels for an additional fee. A trade association in the cable television industry reported that it was not optimistic regarding consumer interest in text-based services, adding that the organization's leaders believed that less than 3 percent of the audience would be willing to pay for text services. A spokesman for the association stated: "I'm not saying the business won't work. But what I'm saying is this isn't the greatest thing since sliced bread."[59]

As newspaper companies involved in cabletext began to accept the market reality, they were forced to rethink the approach to their cable projects. In order to make their cable channels more aesthetically pleasing, newspaper companies began supplementing cabletext with video programming. Some abandoned text-based services in favor of video entirely. In October 1983, as much as half of the newspaper industry's cable television channels included video programming. The ANPA reported that three newspaper cable ventures had closed in 1983, and an association official observed that the "very definite trend" toward video programming "may be the key to success."[60]

An industry executive explained his company's decision to incorporate video programming in the newspaper's cable programming was in response to advertiser concerns. He stated: "Advertisers didn't understand text television. Television to them [advertisers] was sound, sight and motion. Consequently, we decided that we needed some old-fashioned sound, sight and motion laced into the text to keep the cash flow going."[61] Cox Enterprises, which owned newspapers in several markets in the U.S. including Georgia, Florida, Ohio, and Texas, embraced the idea of using video programming on cable to create a television presence for several of its newspapers. Even the *New York Times* announced a deal in September 1983 to provide video programming as part of a joint venture with Warner Communications in which news employees would appear on camera.[62]

The programming provided by the *New York Times* and other larger newspapers may have been relatively well produced, but newspaper companies overall struggled to create video content that measured up to the standards consumers had come to expect from television news organizations. In recounting this period, some observers blamed the newspaper industry's lack of investment in the product that resulted in what was often described as amateurish. In recounting the video efforts of this period, a newspaper executive stated that "newspapers were nickel-and-diming it to death."[63] The newspaper industry struggles with video were especially acute in small markets where resources and talent level did not result in compelling television content.

A publisher of a small daily in Missouri recalled that his newspaper produced weather reports, election news, and local sporting events. "We had all kinds of fun," he said, but added that "We were amateurs. . . . People quickly found out [our newspaper reporters weren't] Walter Cronkite."[64] A Tribune Co. executive later recalled that such video efforts "really didn't take showmanship into account. You can't just put a reporter up there and read."[65] Newspaper companies became uncomfortable with product offerings that were mocked and ridiculed. They had failed to understand that consumers' expectations—even in small markets—were formed by what they experienced from other channels. As newspaper companies realized that their video content would not meet those expectations without significant investment, the industry's collective interest in cable television diminished quickly.

There were, of course, other business issues in play, but in looking back on this time it is clear that newspaper companies soured on the cable business as suddenly as they had earlier embraced it. Newspaper executives who had insisted on long-term deals with cable operators, even though they did not know where the market would lead, decided rather quickly that the market had turned in an unexpected direction requiring more investment than they were willing to make. The following section explores the period when newspaper companies were extricating themselves from cable television projects and discusses some of the rationale provided for the newspaper industry's change of heart.

Newspapers Exit the Cable Market

The newspaper industry's involvement with cable television peaked in the mid-1980s with at least 200 newspapers providing content, including local news and advertising, to cable systems. By the end of the decade, only two such projects remained in operation.[66] Newspaper companies had entered into cable initiatives with extremely high expectations, but they became disenchanted with the prospects for cable television when those expectations were not met quickly. As one official from the industry trade association asserted, newspapers grew less enthusiastic about the prospects for cable ventures because "there aren't any really overwhelming success stories."[67] Market reality and expectations were not aligned, and it resulted in many newspaper companies entering a market for which they were unprepared.

Although newspaper cable activity did not reach its peak until the middle of the decade, 1983 marked a recognized shift in attitude. A newspaper trade association official observed:

> New products and services, like cable, seem to move through distinct cycles of expectations, similar to the business cycle. For cable, 1981–82 were years of unbounded, and unrealistic, optimism. 1983 has brought a much more sober view of cable.[68]

The perspective newspaper executives had about cable television at this time was shaped by the operating realities of their projects, such as the increasing expectations for video programming as noted previously. However, another issue influencing newspaper executives stemmed from cable television's emerging financial structure that worried conservative newspaper managers.

Projections for a typical cable television system in 1982 called for it to make $275 million in profit over the 15-year life of a local franchise agreement. But a cable industry trade association official asserted "that if the . . . revenue projections are as little as 5 percent too high," the cable system "stands to lose even more than that amount."[69] The narrow margin of error was due to the relatively high requirements for capital spending to build and maintain the cable infrastructure. Cable companies were borrowing heavily to fund their capital expansion. By the end of 1982, the cable television industry's collective debt had reached nearly $4 billion.[70] Newspaper managers accustomed to balance sheets with far less debt grew skeptical of the cable industry's long-term ability to produce the profit margins newspaper owners had come to expect from their own operations. As a result, direct investment in cable television systems by newspaper companies peaked at about 16 percent in 1985.[71]

Meanwhile, newspaper companies increasingly were concerned about the operating costs associated with their cable ventures, whether or not they had an ownership position. A trade association official stated: "We want to be realistic about what it costs us; not deceiving ourselves about the business opportunities in this field by not knowing our true costs."[72] To provide its membership with operating guidelines for their cable ventures, the ANPA in 1983 calculated what it considered to be a plausible scenario. In a hypothetical market with 200,000 households and 40,000 cable subscribers, the ANPA said a newspaper cable channel could take in $480,000 in revenue and generate a 27 percent profit. Nevertheless, an association official conceded that newspapers were not at the time generating that level of profit, but instead referred to the numbers as "a worthy and realistic . . . goal" for the next three to five years.[73] As newspaper companies attempted to find their way in this market, several of them followed ANPA advice to reduce resources allocated to information services and increase resources dedicated to generating advertising sales on the cable platform.

The trade association official said the industry group had come to believe "that the business opportunities . . . for a newspaper in cable are largely in the area of advertising," explaining that cable offers newspapers an outlet to "outflank" local television stations, which had emerged as the largest competitors to local market newspapers.[74] In several instances, newspapers decided that the advertising opportunity had nothing to do with their own content. They opted instead to operate as brokers and representatives in local markets for advertising placed on channels carrying such exclusive cable programming as CNN or ESPN, a sports programming channel.[75] These arrangements allowed for local cable operators and newspapers to remain allied against local market broadcast television stations in the battle for advertising dollars.

Newspaper executives saw the expansion of cable television channels as a positive for their own business because they believed that a fragmented television audience made it more difficult for local broadcast television to maintain market share of advertising. An executive explained that an advertiser a few years earlier could reach 90 percent of a typical market through one broadcast buy. That advertiser would struggle to reach 60 percent market coverage in the mid-1980s with only one broadcast buy, the executive said, attributing the difference to the audience fragmentation created by cable's household penetration and channel expansion.[76] Such changes in local advertising markets led an executive with McClatchy Newspapers to state: "Cable television may turn out to be our best friend in the long run."[77] Publishers believed they could position their newspapers with advertisers as a cost-effective way of increasing market coverage in areas of rapid audience fragmentation, thereby capitalizing on the market turmoil cable television had created. So despite the shifting relationship with cable television operators, newspaper companies continued to believe that cable television was more friend than foe.

Advertising representation and brokerage deals may have salvaged some of the business relationships between cable television operators and newspaper companies, but those arrangements were far afield from the original vision of newspapers as the source of robust information services. In announcing the closure of its cable venture, the *Florida Times-Union* in Jacksonville, Florida, was among the first newspaper companies to address publicly some of the shortcomings that prevented it from realizing the vision that had originally attracted the newspaper to cable television. The newspaper closed its cable channel after only three years of operation, blaming the failure on the passive nature of cabletext and various technical shortcomings.

The *Florida Times-Union* was an example of a newspaper that began with cabletext before supplementing its content with video-based news reports. An executive maintained that the simple cabletext service suffered from a lack of interactivity, making it unattractive "for classified advertising because it would require viewers to watch the screen until the advertising in which they were interested in scrolled by."[78] The move to video failed to attract a viable audience in part due to unresolved technology issues, he said. For example, the executive said signal interference from a local broadcast station impaired the visual quality of the newspaper's channel.[79] As their interest in the cable television platform waned, newspaper companies had no shortage of reasons for explaining why the ventures had failed to meet the lofty goals envisioned only a few years earlier. From a timing perspective, however, the move away from cable in the mid-1980s coincided with the high-profile closings of Knight-Ridder's Viewtron service and Times Mirror's Gateway project.

The proud newspaper industry had shifted backwards, from leader to laggard, in the marketplace for electronic distribution of media content. As an industry, its efforts to exploit electronic distribution were not keeping pace with overall market developments. But where the rhetoric surrounding those closed videotext projects had been tinged with failure, the newspaper industry positioned its exit from cable television as a strategic withdrawal. The pullback from cable television was described as a shift in how newspaper companies chose to allocate resources rather than as the failure of expectations that it actually was.

Positioning the Exit

The economic value of cable television systems soared in the mid-1980s as television viewers became more accustomed to receiving their programming through a wire rather than an antenna. By the end of 1985, cable television was available to nearly 73 percent of homes with televisions.[80] The gold-rush description discussed earlier had proven to be an appropriate analogy, given that cable television systems valued at $200 to $300 per subscriber in the mid-1970s sold for $1,100 to $1,200 per subscriber in 1986.[81]

Many newspaper companies that had invested in cable television in the early 1980s were ready to cash in on those investments by 1986. "I think our timing has been rather fortuitous," said a McClatchy Newspapers executive in referring to the escalation in value of cable television systems.[82] McClatchy Newspapers is representative of its industry's mid-decade retrenchment. "We flirted with diversification and decided we do better at newspapers," said another one of its executives.[83] As newspaper companies across the board sold off their cable ownership holdings or negotiated their way out of channel leases, executives echoed those sentiments and talked about focusing on their core product—printed newspapers.

One industry analyst stated: "Newspaper companies are saying newspapers are going to be the backbone."[84] The change in strategy, which downplayed new ventures, was so significant that the industry's own trade association publication, *Presstime*, published a special report about it. It stated in part:

> McClatchy's narrowing focus reflects what observers detect as a recent trend in the newspaper business: that following a period in which they aggressively entered businesses not directly related to their traditional mission of disseminating news and advertising, many companies are going "back to basics." In general, they are re-emphasizing their role as print publishers—and, to a lesser extent, radio/TV broadcaster—and de-emphasizing their involvement in other enterprises.[85]

The "back to basics" rhetoric was supported by other industry activity. For example, newspaper companies reduced their investments in the radio business as well. From 1982 to 1985, the newspaper industry reduced its ownership of the country's AM radio stations from 23 percent to 5.2 percent, while its investments in FM stations declined from 37 percent of the total to 6 percent.[86]

Tribune Co. sold its cable holdings, but kept most of its broadcast properties, while Dow Jones, Harte-Hanks, and Gannett are also examples of newspaper companies that either abandoned cable television or significantly reduced their stake in the industry during this period.[87] Other newspaper companies such as Times Mirror and Cox Enterprises separated their cable television holdings and newspaper holdings into distinct subsidiaries, recognizing that the capital spending requirements of cable television was too great to share a balance sheet with newspaper activities. But for some enterprises, creating separate businesses under a common holding company did not go far enough.

Executives with the Tribune Co., for example, said they made the strategic decision to withdraw from cable entirely because the capital required to become a major player in the industry was prohibitive given their desire to remain committed to the newspaper business and to expand its broadcast holdings. It would have required "much more than we wanted to spend," according to a company spokesman.[88] Although newspaper companies were extremely profitable during this period relative to other businesses, the amount of investment required by the cable industry illustrated that the conservative newspaper business had its limits. Analyst comments from this period suggest that such limitations stemmed from owners who were reluctant to take on risks that could jeopardize profitability. Even at larger media conglomerates, newspapers were responsible for generating the most revenue and profit, and executives had little incentive to put the newspaper franchise at risk.[89]

Taking advantage of the escalation in cable values, while at the same time positioning the cable retreat as a return to core products, resonated positively for the newspaper industry in the Wall Street investment community. Analysts believed that shareholder pressure to maintain the industry's high profit margins contributed to the decisions to curtail ventures outside of the core product. This was especially true of the publicly traded companies such as Gannett and Times Mirror.[90] An important aspect of this discussion, however, is that by characterizing the retreat from cable as a re-deployment of assets, newspaper companies were allowed to downplay what was essentially failed execution. Newspapers had rushed into the cable television business without appreciating the complexities of the market and with no consumer research to support claims that a market for text-based information services on television could be developed.

In looking back over this period, a Tribune executive said newspaper companies acted impulsively, seeking "franchise protection [rather than] franchise extension."[91] And by acting impulsively, newspaper companies had not done enough diligence to be prepared for shifts in the market that would be inevitable—the requirement to add video programming to their news channels, for example. A McClatchy Newspapers executive said that newspaper companies had underestimated what would be required to operate a cable television channel, stating that "it turned out not to be as easy to run" as newspapers had assumed it would be.[92] An article in *Presstime* summarized the issue, stating that newspaper companies had concluded that "cable television and other endeavors simply turned out to be more trouble than they were worth."[93] These comments underscore a line of thinking that newspaper companies were never invested in leading an electronic revolution, only in defending their local markets against one.

As was discussed in Chapter 1, the demise of several high-profile videotext projects left the industry scarred by the experience. The aftermath of those projects affected how newspaper companies approached subsequent online initiatives, including the Internet era, which will be discussed in a later chapter. However, the newspaper industry's involvement with cable television was not cast in the same light. Some observers even felt the newspaper industry had been bolstered rather than beaten down by its involvement with cable television.[94] Wrote Patten:

> Newspaper ventures into cable ownership were harmless and in some cases instructive. They served to increase awareness of a changing media environment. And the newspapers learned a few lessons about the problems of profitability in a new communications field.[95]

This more positive reaction stemmed, at least partially, from cable television's position within the media marketplace. Cable television had come to be viewed differently than other existing media. Finnegan and Viswanath asserted that "cable is not a medium *per se*, but rather a delivery system of channels of varied content."[96] Therefore, newspaper companies were seen as backing away from a distribution platform rather than a full-fledged medium, which made strategic sense within the business and investment community.

The newspaper industry experience during this period also fostered a sense that the competitive effects of cable television had been overstated. One executive stated: "As it turns out, cable was not nearly as great a threat to newspapers as some people once thought it might be."[97] Another executive said that the initial interest newspaper companies had in cable television emanated from the belief that such systems would be a major distribution platform for newspaper content, but by the mid-1980s, he said: "it's become more and more apparent any impact cable is going to have on newspapers is a lot less than anybody imagined."[98] This perception that the newspaper industry had dodged a significant competitive threat from cable television was repeated by analysts who followed the industry for investment research.

These analysts generally concurred that the industry's renewed emphasis on its core business was the correct strategy in 1987. One investment banker proclaimed that newspaper companies did not need to worry about investing in new endeavors because there was "no new media coming along" that promised them a better use for their capital.[99] Another investment analyst proclaimed: "we can see that newspapers will survive that [competitive] onslaught."[100] The newspaper industry was emboldened by the belief that its business model, largely dependent on a local market newspaper monopoly, was well suited to fend off threats from newcomers. The economic fundamentals of daily newspapers posed a huge barrier to entry for competitors. While radio and television (including cable systems) had made inroads in local market advertising over the decades, the newspaper model had not cracked. Daily newspapers remained hugely profitable and were at the forefront of the local media economy.

Newspaper companies also began to assume that audience fragmentation brought on by cable television and other forms of technology-based media would benefit them over time. As one executive for Tribune Co. surmised, the local newspaper franchise would flourish as this fragmented audience sought an information "starting point" and "needs a single, reliable resource."[101] Even the physical form of the newspaper received newfound respect in the mid-1980s as cabletext was deemed too passive and videotext too slow and plodding. Said one newspaper broker: "the portable, clippable smorgasbord that [readers] get in their daily newspaper is a cost-effective package that will be very hard to improve upon."[102] The juxtaposition of these more positive views with those expressing failure in the wake of Viewtron's demise reflect the dichotomic nature of the newspaper business in the 1980s.

Discussion

Newspaper companies were indeed successful businesses and the industry had weathered numerous competitive battles, but concern about the risk from new technologies influenced executive action more than any other business fundamental during this era. This led newspaper companies to embrace opportunities in the cable television market at the beginning of the 1980s with the same initial exuberance they had demonstrated for videotext projects during this period.

A significant underlying influence on the decisions to invest in cable television—although rarely addressed straightforwardly by newspaper industry executives—stemmed from the concern that emerging technology would alter a business model that, as noted earlier, had come to rely extensively on local monopoly control. Gomery, for example, asserted in a media economic analysis that "some sort of dramatic technological breakthrough" would be the most likely occurrence that would break the newspaper industry's local market monopoly.[103] Newspaper industry leaders in the 1980s wanted to prevent that from happening. They wanted to own such a breakthrough and were exploring emerging technologies to determine how effective they would be in allowing the industry to gain an even stronger position in local markets. They specifically were interested in regaining a portion of the advertising revenue that had gone to local market television and radio as the electronic era took hold, first in the 1930s with radio, and later in the 1950s and 1960s with television.

When direct investments by newspaper companies in the emerging technologies did not provide the immediate payback that many newspaper executives expected, however, they were quick to withdraw from what they saw as expensive experiments. Many became comfortable with the notion that electronic distribution platforms had arrived, but were not close to supplanting the superiority of the printed newspaper. The prevalent industry sentiment was that since newspapers had not succeeded in either videotext or cable television, technology-based challenges were a distant threat at the very least. If newspaper companies themselves could not make a sustainable business case for videotext or cable television, this line of thinking went, then it was unlikely competitors would be successful in using those platforms against newspapers.

Nevertheless, the potential for what electronic distribution could become haunted the industry even as its members dismantled their electronic ventures. As a result of these ongoing concerns, the newspaper industry became fixated on making sure that if it did not exploit the potential of online media, them no one else would, either. The industry became especially adamant that a breakthrough in online media would not come from the telecommunications industry, which was also a beneficiary of local market monopolies. Although newspaper companies had partnered with cable television operators, as discussed in this chapter, and had worked closely with telecommunications companies on videotext projects, as was explored in Chapter 1, the aftermath of both sets of activities resulted in a more insular newspaper industry.

Newspaper companies had no interest in partnering with the telecommunications industry as they had done with cable television. Instead, concern over the potential competition from the telecommunications industry escalated into public hostility and became a significant influence on the newspaper industry's activities in the 1980s. Despite the level of activity chronicled in this chapter and the previous chapter, it can be argued that the newspaper industry's most determined effort during the 1980s was the lobbying mustered in defense of its local markets against the telecommunications industry. Chapter 3 examines the story of this effort and the heated political battle that ensued.

3

Newspapers React to Fear of Telecommunication Dominance

AT&T—then the regulated monopoly telephone service provider in the U.S.—conducted a "concept trial" in 1979 for a service that would allow customers to "call a database to retrieve directory listings and other information such as sports results, time and weather."[1] The trial was held only in Albany, New York, and AT&T mentioned the project only briefly in its annual report to shareholders released in the spring of the following year. The description was rather innocuous, noting that "information was displayed on a TV-like screen" and that more tests of the system were planned.[2] This test received little press coverage and most consumers were unaware of it. However, it shook the newspaper industry's executive ranks, contributing to "The AT&T 'scare.'"[3] Newspaper publishers feared regulators would free AT&T to "become a major information provider in the pioneering days of various new telecommunications technologies."[4]

This chapter examines how the newspaper industry reacted to the possibility of direct electronic competition from the nation's telecommunications industry. Newspaper publishers already viewed the phone companies' Yellow Pages directories as advertising competition. However, the prospect of AT&T offering those vast encyclopedic listings through online databases seemed like an unfair advantage from the perspective of newspaper publishers. This chapter explains how the newspaper industry sought to derail that perceived threat, and in doing so, provides insight into the newspaper industry's evolving perspective of its market and competition in the early years of online media. The newspaper industry's lobbying campaign was highly successful in thwarting many of the telecommunications industry's plans in the 1980s and early 1990s, but those results came with a price. The industry, once seen as a technology pioneer, developed a reputation as being defensive and protectionistic, interested only in protecting its local market monopoly rather than fostering innovation.

A brief summary of the telecommunications industry regulatory environment that existed during the early 1980s opens the chapter to provide context for explaining the newspaper industry's position. The chapter then explores the intense rhetoric that emanated from the newspaper industry's lobbying effort, which was designed to influence regulation that would restrict how the telecommunications industry could participate in the electronic information services market. The breakup of AT&T is discussed as a milestone event, including its aftermath when the regional Bell operating companies emerged as powerful forces in the telecommunications industry and emerged as new forms of competition for the newspaper industry.

Chapter 1 examined the newspaper industry's early videotext projects, while Chapter 2 explored the industry's investments in cable television. The decisions recounted in those chapters—to close Viewtron and similar electronic endeavors and to abandon cable television investments—combined with the regulatory stance against the telecommunication industry to frame the newspaper business as an industry in retreat. By the late 1980s, without any clear victories in the electronic realm, the earlier optimism that newspaper companies would lead the way in online media faded. Therefore, the events and issues discussed in this chapter are an important bridge in understanding how the newspaper industry evolved as it did in the years leading up to the Internet.

Regulatory Background

Due to several long-standing consent decrees and regulatory rulings by the U.S. Justice Department and the FCC, the nation's telephone system controlled by AT&T was not allowed to expand into businesses outside the scope of its status as a common carrier of communications services.[5] The effects of those restrictions were debated because many viewed the telephone wire leading into homes and businesses as the natural way to access the array of electronic information services that were looming on the horizon. Keeping the owner of those wires from profiting from such information services seemed counterproductive to those who believed the restrictions would stifle innovation. Others, however, believed AT&T and its Bell operating companies would use monopoly power to thwart competitive services, which would also undermine innovation.

The emergence of cable television during this period was a major factor in shaping the debate surrounding the future of the country's telecommunications infrastructure. As cable television expanded, it became another wire into the home and represented the possibility of real competition for the telecommunications industry in the delivery of information services. As Noam wrote in 1982:

> The entry of cable television into the American household was not planned as part of an alternative telecommunications system. But now that it is becoming a fact, one should make the most of it. . . . It will not be feasible to contain the possibilities of the technology and to deny their services to consumers. If technology is destiny, it spells out a future of integrated telecommunications.[6]

Noam's observation that cable television's influence in the market was an unplanned event underscores how the spread of technology in the early 1980s confounded regulators and policymakers.[7] It was the time when the clear boundaries that separated television and telecommunications began to disintegrate, and new ideas about information services began to take shape. How the country should regulate within this new environment—especially as it concerned the future of AT&T—became one of the most important policy issues of this era.

The AT&T Conundrum

The most fundamental decision confronting the U.S. government in the 1980s regarding its policies toward media technology and information services related to the course of action to take with AT&T. One option was to preclude AT&T from directly participating in information services "on the grounds that a common carrier may not simultaneously act as a processor of data for public service."[8] Another option was to follow the lead of several Western Europe democracies, which had granted extensive authority to their telecommunications and postal monopolies to introduce new electronic information services. Smith noted that if the U.S. followed that model, it could "make AT&T the focal point of a vast expansion into the role of national data storage and disseminator."[9] Any change enacted by the government would alter the telecommunications market that had operated under a consistent set of rules for more than two decades.

AT&T and the federal government had entered into an antitrust consent decree in 1956 that defined AT&T as a common carrier and restricted it from providing information services directly to consumers. By the 1980s, however, Congress and the FCC understood that changing technology would make it necessary to revisit the nation's telecommunication policies. In early 1980, for example, the FCC proposed rule changes to allow AT&T to offer data retrieval services through a subsidiary.[10] It was proposals such as this that raised the ire of the newspaper industry, which argued that a regulated monopoly had an unfair advantage in the marketplace and that it would use that power to keep competition from developing. The argument that a separate subsidiary would create transparency and prevent AT&T from leveraging its telephone monopoly to its advantage in information services held little sway with newspaper publishers.

This chapter is not intended to recount all of the regulatory and political machinations that led to a new consent decree that broke up the AT&T monopoly. Rather, its purpose is to explore how the newspaper industry reacted to the shifting telecommunications landscape that led to the breakup and paved the way for a major rewrite of the national telecommunications law. During this period, the newspaper industry became a powerful lobbying force in the public debate over telecommunications policy. The newspaper industry's actions relating to telecommunications policy shaped its approach to the information marketplace and altered its public image during this period. The next section explains AT&T's positions that provoked the newspaper industry's lobbying response.

AT&T's Stance

Regardless of how regulations were changed, telecommunications industry observers were convinced that AT&T's presence in the electronic marketplace would be huge. Given the monopoly's power and position, it would have been unfathomable in the early 1980s to think the government would prevent AT&T from playing a leading role in developing information services. After all, the research and development capabilities of the company's Bell Laboratories unit had achieved legendary status, and AT&T executives anticipated with excitement the possibilities of the emerging information marketplace.

In the letter to shareholders included in the AT&T 1980 Annual Report, company chairman Charles Brown wrote that "no longer do we perceive that our business will be limited to telephony or, for that matter, telecommunications. Ours is the business of information handling, the knowledge business." He stated that "it appears to be widely if not universally agreed that—in an era of intensifying competition—it no longer makes sense to deny the Bell Systems the opportunity to compete in unregulated markets." He concluded, "The technology of the

Information Age is ours. Indeed it was Bell System technology that very largely brought it into being. And it is Bell System technology that positions us to fulfill its opportunities."[11]

Pontificating about the potential of new information services markets was easy, but Brown understood that capitalizing on those opportunities would require disarming potential adversaries and influencing key politicians and regulators. Brown knew that AT&T's critics included the powerful newspaper industry, which prompted him to accept an invitation to address a gathering of influential newspaper executives. But rather than win over any converts from the ranks of the newspaper industry, the meeting is recognized as the opening salvo in a war of rhetoric.

Newspapers, AT&T Launch War of Rhetoric

When AT&T's Brown stepped to the podium during the meeting of the newspaper trade association's telecommunications committee in early 1980, he initially sought to calm a contentious atmosphere. Brown told the committee that the newspaper industry's fears were unfounded and that publishers were "seeing ghosts under the bed." He elaborated: "If what you're concerned with is, 'Are we going to provide a news bank?' the answer is no. We're not interested in that." He said that AT&T had no plans to field its own newsgathering operation and that concerns over the company's intentions were overblown. "I think you've reached quite a long way if you think we're interfering with the freedom of the press," he said.[12] However, if Brown's appearance before the publishers' committee was designed to appease the newspaper industry as AT&T sought reductions in its regulatory restrictions, the strategy did not work. During his remarks, Brown suggested that "a philosophical difference" existed if the newspaper industry felt that the restrictions on information dissemination placed on AT&T extended to such content as weather and sports scores.[13] AT&T viewed such material as commodity data rather than news because it required no editorial judgment to collect and distribute. It was an early admission that AT&T was, in fact, interested in collecting and distributing information that newspaper companies felt went beyond its charter. But the difference—philosophical or otherwise—regarding the definition of news content was not the focus of the disagreement. The newspaper industry was more alarmed by AT&T's plans in the area of advertising services.

Given the phone company's enormous directory publishing business at that time, newspapers felt especially threatened by the possibility of those directory listings being used to populate online services that would compete directly with newspaper classified advertising. Brown was asked whether AT&T planned to enter the classified advertising arena through online services. Brown said AT&T did not want to be "excluded" from services that relied on technologies AT&T was actively developing.[14] It was not the response the newspaper industry wanted to hear. Brown's remarks—candid and delivered in person—galvanized the newspaper industry and unified its leadership in opposition to AT&T's effort to reduce regulatory restrictions.

The Newspaper Industry Responds

Before the meeting with Brown, newspaper publishers conceptualized information services in terms of some future technology. Following the meeting, however, newspaper publishers were more concerned about the immediate ramifications of AT&T's actions. Brown's comments had the unintended consequences of turning AT&T into a tangible threat. The general counsel for the newspaper industry's trade association later mockingly referred to the event as the time "Charlie Brown came to dinner," but seriously added how Brown's appearance changed the newspaper industry's thinking about "the future electronic information marketplace."[15] He wrote:

> What became obvious . . . was that while AT&T said it had no plans to hire its own news staff, it clearly had designs on the electronic publishing of the future. This included aspects of the business in which AT&T's control of monopoly telephone services and facilities could pose severe anticompetitive threats to future electronic publishing competitors.[16]

The newspaper industry's leading attorney recalled Brown's presentation as the impetus for setting in motion the trade association's unprecedented lobbying effort "to try and modify and, if necessary, oppose legislation . . . that would have given the green light to AT&T's electronic publishing plans."[17] Despite the newspaper industry's own hefty revenue and profits during this period, its first salvo in the lobbying effort portrayed AT&T as an even larger enterprise that could not be trusted to grow larger through new information services businesses.

Soon after the meeting with Brown, the chairman of the newspaper industry's telecommunications committee issued a statement that summarized the industry's position and called for Congressional action to prevent the FCC from easing the competitive restrictions on AT&T:

AT&T has a revenue base that is larger than the sales of the newspaper, television and radio industries combined. Under federal protection, this giant company has developed an electronic distribution network that reaches more than 90 percent of the homes in its markets. The action by the FCC raises serious, unanswered questions concerning the ability of newspapers to be able to compete fairly in this environment. This focuses even greater attention on just how the Congress will deal with this issue.[18]

The newspaper industry's leadership was adamant in its rhetoric that an unleashed AT&T would be detrimental to the country's free press. However, the industry's collective decision to intervene in the legislative and regulatory rulemaking process through a lobbying effort carried out by its trade association was a radical departure in tactics. And, it was an effort not without critics.

Fink recounted this period as a time when "incredibly, the newspaper industry blindly closed ranks" and "lobbied Congress and the public to keep telephone companies out of the information business."[19] Fink asserted that the newspaper industry had taken a position that ran counter to its historical role of supporting freedom of expression. He wrote:

Newspapers, which since Colonial times demanded the right to free expression, positioned themselves in public perception as arguing that telephone companies should be denied that right. No industry in modern times has made a worse strategic error. Newspaper executives somehow decided they could lobby away new competitors and, with them, a new, exciting technology. It was as if horse ranchers and carriage makers had lobbied Congress to keep Henry Ford from building automobiles.[20]

The following section explores the issues that caused the newspaper industry to react so passionately, and explains how that reaction resulted in altering decades of industry behavior as the industry's lobbying campaign unfolded.

The Newspaper Industry and Political Lobbying

As proposed legislation was debated in the Congress regarding the proper role for AT&T, the FCC agreed to a further review of its proposed regulatory changes. Therefore, the issue was active both in the legislative arena of Congress and in the regulatory forum of the FCC. The newspaper industry's lobbying effort was aimed at both fronts and took on the intensity of a political campaign as industry leaders urged lawmakers to refrain from enacting legislation hastily that would affect telecommunications policy for decades to come. They also urged regulators to take a holistic view of the marketplace when deciding how to implement rule changes.

The executive leader of the newspaper industry's trade association framed his group's opposition to AT&T in the form of a question he posed to lawmakers and regulators:

> Should the nation's largest company, AT&T, which has grown and operated under specially granted monopoly privileges, be permitted on any scale to involve itself actively not just in the common-carrier transmission of information but also in the selection, editing and vending of that information to the public in a mass-media or data-retrieval sort of structure which in the United States raises the most basic sorts of social, constitutional and anti-trust concerns?[21]

The approach represented by such statements was risky politics for the newspaper industry. By discussing the size and scope of the telecommunications industry, the newspaper industry risked having its own profit levels scrutinized by the government. In discussing monopoly power, the newspaper industry also risked comparisons to its own position in local markets where a single newspaper company often dominated the advertising market during this period. Furthermore, by raising the spectre of antitrust concerns, the newspaper industry risked legislative review of its own antitrust exemptions that allowed several companies to operate newspapers under joint operating agreements.[22]

The willingness to open the debate on so many fronts underscored how serious the newspaper industry believed the threat from AT&T to be; but as noted, the act of getting so involved in the law-making process marked a dramatic change for the newspaper industry. LeGates observed that prior to this period "the newspaper business has enjoyed a kind of moral aloofness from lobbying or pressuring the government on its own behalf"; but he added, "We see this era drawing to a close."[23] LeGates wrote that such involvement stemmed from the newspaper industry's recognition that its marketplace was changing:

> ANPA last year chose to intervene in the legislative process, not on a bill focused on the newspaper industry, but on one . . . whose intended thrust was telephone deregulation. This was but one of a string of interactions and confrontations with parts of the information industry newspapers never had to worry about before. That time is gone; newspapers today must adjust the perspective from which they long have viewed the information world.[24]

It is important to understand that within the timeline of this book, the lobbying effort against the telecommunications industry's entrance into information services occurred simultaneously with the newspaper industry's own forays into videotext and cable television ventures that were discussed in the first two chapters. Within this context, the initial lobbying effort also can be viewed as simply another early reaction to the changing marketplace, another attempt to regain footing in the shifting sands of technology change. By holding the telecommunications industry at bay, newspaper companies believed they would have more time to develop their own presence in electronic information services as the market evolved.

In any case, the newspaper industry's actions on so many fronts contributed to confusion about its position and ultimate motives. In the stance against AT&T, the newspaper industry argued that electronic services were a threat, as LeGates observed. However, the newspaper industry's own investments in numerous early electronic information services illustrated how they "may be . . . not a threat but an opportunity."[25] The newspaper industry's own relationship with the telecommunications industry as a customer—in some cases availing itself to controversial discounted rates—further complicated the rhetoric.[26] Even as newspaper industry lobbyists were ramping up their war of words against AT&T, companies within the industry were extending their ties with the telecommunications company. As was discussed in Chapter 1, for example, AT&T was Knight-Ridder's technology partner in launching the Viewtron service. These arrangements between telecommunications companies and newspaper companies during this period demonstrate the complexity of the relationships and show that the exchanges were not always adversarial.

Even though much of the newspaper industry's early rhetoric was cast in the semantics of keeping the flow of news and information out of the hands of a giant regulated monopoly, protecting advertising revenue was a central part of the newspaper industry's agenda. Indeed, some 15 years later, the newspaper industry suffered deep erosions in its advertising revenue as it confronted numerous online competitors such Monster and Google, which underscores how the industry's concerns about advertising revenue were well placed. This issue will be discussed in later chapters as well, given that advertising always played an important factor in how newspaper companies approached online media. At this juncture, however, newspapers were hearing dire warnings about the potential competition. LeGates, for example, stated that "One of the major challenges facing newspapers in the coming years will be to preserve the income stream from advertising." LeGates specifically discussed the threat to classified advertising—which he called "one of the mainstays of the newspaper income stream"—posed by the emerging telecommunications and computer technologies:

This advertising is for all practical purposes a data base service, albeit one that is offered on paper. Other companies could easily offer it by computer. The advantages of computer readable classified advertising are quite convincing. It could be up-to-the-minute, searchable by the reader and contain other properties that may be indexed.[27]

Nevertheless, the newspaper industry's position refrained—for the most part—from discussing the advertising issues outright. Rather, the lobbying effort put forth a position that became known as "the diversity principle."[28]

Newspaper industry leaders argued that if AT&T was allowed to enter the information services business unfettered, its control of the network over which the information flowed would lead to a reduction in competition because of the inherent advantages AT&T would have in the market. A newspaper industry lawyer created the phrase "diversity principle" to explain the industry's view that there was a "need to separate content from conduit in emerging, telephone-based, electronic information systems."[29]

Newspaper leaders described the information marketplace in the early 1980s as extremely diverse, with more than 1,700 daily newspapers, 7,500 non-daily publications, and more than 10,000 magazines. They claimed that such diversity of titles meant that competition was alive and well in the printed information market, but that landscape would be diminished "when AT&T, the world's largest corporation, controlling over 80 percent of the telephones, decides to become an information provider over its own local exclusive distribution system."[30] The newspaper industry's position was simple: Congress had to ensure that AT&T remained regulated at least to the extent to which it could use its own network for the dissemination of information services.[31]

The newspaper industry's position attracted critics. Kinsley, for example, accused newspaper publishers of trying to protect their own local monopolies by keeping a national monopoly out of the business. He cited this statistic: "Of 1,560 American cities with newspapers, only 34 have true newspaper competition."[32] Kinsley's position was that the newspaper industry argument was not altruistic, but rather self-serving. He wrote: "The publishers don't really fear an AT&T monopoly on the news. What they really fear is losing their own monopoly on local advertising."[33] Kinsley was critical of newspapers for using their own editorial pages to espouse the industry agenda. He said the industry's position contained "complete inversions of meaning," and concluded:

> The publishers want government restrictions on market entry, and call it promoting "diversity." They want to protect their monopoly in the name of "competition."

They talk about the "marketplace of ideas" when they really mean the marketplace of advertising, and they warn of the "peril" of restricting commercial speech when they really want to restrict commercial *and* noncommercial speech.[34]

Despite such criticism, the lobbying effort was deemed successful by newspaper industry leaders who believed their intervention prevented legislation from passing in the early 1980s. Moreover, the diversity principle, which the newspaper industry created, became an important tenet within the country's ongoing telecommunications policy debate. With Congress lacking the votes to pass sweeping telecommunications reform in the early 1980s, the fate of AT&T was left to the federal courts overseeing an antitrust suit brought by the government against the AT&T phone monopoly.

The Breakup of AT&T

In January 1982—about a year after an antitrust trial had begun—AT&T and the U.S. government entered into a new consent decree designed to settle the case. It was a sweeping agreement to break up the monopoly phone system into a "new wholly competitive AT&T" and a collection of regulated Bell operating companies that would be owned by seven regional holding companies.[35] The target date for the breakup was set for January 1, 1984. The agreement gave AT&T the freedom it had sought to explore some new telecommunications markets, but it barred the new AT&T from electronic publishing for at least seven years to give competition time to become established in what was considered a nascent market.[36] The court also allowed the regional Bell companies to offer information gateway services and to provide a unified billing methodology for them. In subsequent rulings a year later, the court said the regional Bell companies also could provide other information services such as voice storage and retrieval services as long as they did not include "content generation or content manipulation."[37]

The newspaper industry trumpeted the AT&T breakup as a "victory" and applauded the ruling by Judge Harold C. Greene, which prevented AT&T's immediate entry into the broad electronic publishing industry on grounds that it could undermine the "First Amendment principle of diversity."[38] The judge stated: "AT&T's mere presence in the electronic publishing area would be likely to deter other potential competitors from even entering the market." He added, "AT&T's ability to use its control of the interexchange network to reduce or eliminate competition in the electronic publishing industry is the source of this threat."[39] While the fine print of Judge Greene's order defined for the newspaper industry what would be considered commodity information going forward, the industry reveled in the notion that the diversity principle it put forth had become, in essence, the "law of the land."[40] AT&T was allowed to remain a provider of basic directory listings of name, addresses, and phone numbers as well as recorded time and weather information, but the newspaper industry had achieved a seven-year window in which to exploit the online information without competition from the telecommunications industry.[41]

During the time leading up to the historic consent decree, AT&T's Brown had returned to the newspaper industry association again as a featured speaker. A trade journal's account of the appearance said Brown "expressed amazement at the extent of the negative attitude toward AT&T within the newspaper industry" when he acknowledged that he had been identified "only half-jokingly . . . as the enemy."[42] The overall nature of Brown's remarks, however, was conciliatory. He told his audience that while a revamped AT&T did not want to be precluded from the electronic "Yellow Pages" business, the ownership structure supporting that business was less important. "We don't care who owns the data base. We make our money on the transmission," he said. He called proposed legislation to bar his company from the business of information services "protectionism," but added that he had grown "weary" and had "no stomach for argument over turf."[43] In a nod to the newspaper industry's successful lobbying effort, he told the publishers they "can get pretty much whatever you want" in terms of new legislation, and encouraged them to think about how AT&T could emerge as a potential partner:

> It is left largely to your own judgment as to how far you want to go to build fences around your primary communications supplier. . . . It is evident to me that what my business, telecommunications, and your business, publishing, should be debating is the prospect for collaboration.[44]

In retrospect, Brown's remarks foreshadowed his company's decision to sign the new consent decree a few months later. AT&T executives understood the political climate made it unlikely for them to win any major concessions in Congress. The consent decree, therefore, became the most plausible way for the company to remove the cloud of the antitrust suit in order to move forward.

The Aftermath of the Decree

Initially, newspaper executives touted the virtues of the decree as fair to all parties involved. Marbut stated, "Nobody walked away with all the marbles, but everybody walked away with some of what they wanted for their own special interests."[45] However, as details emerged regarding the practical application of the consent decree, it became clear that AT&T would have more opportunity to compete in the electronic marketplace than was originally thought to be the case by the newspaper industry. As Mowshowitz stated: "AT&T's newly won freedom to enter unregulated markets puts it in a position to compete with the publishing industry in providing information services."[46] As this became more and more evident during the two-year period between the announcement of the new consent decree and the actual breakup of AT&T, the newspaper industry's language of "victory" turned more cautious with phrases such as "mixed blessing" and "tone of uncertainty" creeping into industry articles about the unfolding events.[47]

On the one hand, newspaper companies began to view AT&T as a potential customer for information services they could create. On the other, as the terms of the consent decree became more widely understood, newspapers began to feel that the competitive threat from AT&T was far from neutralized. While the agreement precluded—until at least 1989—AT&T from offering electronic information services over the long-distance lines it would still control, there were no restrictions that prevented it from creating services that could use the lines of its new children, the regional Bell operating companies. There also was growing concern among publishers that their phone service rates were about to soar in a deregulated environment.[48]

Nevertheless, on the eve of the breakup, AT&T was engaged in numerous partnership discussions with a variety of newspaper companies and said it was optimistic that arrangements could be made for newspapers to provide AT&T with editing and newsgathering services as it expanded its product offerings.[49] As one media consultant put it, "Deregulation has set in motion a series of events that newspapers can see either as a threat or opportunity."[50] When viewed in those terms, the newspaper industry was in the exact same position following the breakup of AT&T as it was before: pondering how to react to perceived threats while at the same time considering how to exploit the opportunities arising from the changing marketplace.

A Period of Complacency

As 1984 came to a close—the end of the first year of AT&T's breakup—the newspaper industry decided that it was "quite clear that electronic publishing will not evolve as quickly as once predicted," wrote Criner and Wilson, both telecommunications policy officials for ANPA.[51] Many issues remained far from settled, given the expectations that AT&T and the regional Bell companies would be petitioning the court for permission to penetrate the information services in ways the original agreement prevented. Nevertheless, Criner and Wilson wrote that due to the pace of change occurring slower than anticipated, "interest has cooled, some publishers have grown complacent."[52] In a follow-up essay assessing the newspaper industry's position at mid-decade, Criner was outwardly critical of the industry. She acknowledged the industry had experimented in a number of information services ventures "with mixed results," but she stated that "newspapers have relinquished their leadership role in electronic publishing."[53] She concluded the essay with a warning to her association's membership:

> Newspapers that have adopted a "wait and see" posture toward electronic publishing may find someday that entry costs and the learning curve are steeper and longer than anticipated. . . . Telecommunications equals competition. As newspapers reflect on the next five years, they must look beyond their traditional industry boundaries at a host of competitors who are developing new products and services. In many cases, those products won't compete directly with newspapers, but they may begin to nibble away at newspaper revenues. . . .[54]

Despite such cajoling from the leading trade association, newspaper publishers remained cautious and did not rush to back any significant electronic services initiatives. As noted in the previous chapters, newspapers around this same time had retreated from their cable television experiments and were winding down several videotext initiatives. The overall decline in electronic services activity reflected the refocus on the core printed product that was discussed in Chapter 2. However, when this retrenchment is juxtaposed with the lobbying against the telecommunications industry, the newspaper industry came across, according to Fink, as "self-serving, 20th century Luddites."[55]

Nevertheless, newspapers began to deploy a wide array of voice services offered over the telephone in what became known as audiotex.[56] Several newspapers, such as the *Houston Chronicle*, offered information including sports scores, stock quotes, and weather, through a phone service as part of a promotional effort without the intention of making money. Other voice projects were intended to be money-makers, but one of the more ambitious examples undertaken by the *Los Angeles Times* was shut down when expectations were not met after only seven months in operation.[57] The ANPA reported that less than 20 newspaper companies were operating voice information services in 1987, which led one executive to observe that most newspapers were merely "dabbling" with such technology.[58] The number of papers involved with such services grew over the latter half of the decade. At least 75 papers were offering free audio services, while about 600— nearly a third of all newspapers—were selling paid content through phone-based systems by 1991.[59] Newspaper executives viewed such projects as low-cost ways to explore alternate methods of information delivery, but they were never embraced by large numbers of a public waiting for the promises of the Information Society rhetoric.

A Period of Wary Cooperation

The voice-based projects initially fostered an increased dialog between the telecommunications and newspaper industries in the mid-1980s. Executives from both industries held a series of meetings in an attempt "to cultivate common ground and nurture alliances."[60] In the summer of 1985, representatives of various newspapers and telephone companies announced a goal of creating "an efficient national videotex network in which the Bell companies supply the transmission lines and newspaper-owning companies provide information."[61] Such a national platform never materialized, but the discussion of it contributed to pockets of regional cooperation. Once the Bell companies were cleared to provide information gateway services, several newspaper companies pursued partnerships with them. In one example, Cox Enterprises, the publisher of the *Atlanta Journal-Constitution*, launched an information service in connection with BellSouth, then the regional Bell company responsible for providing telephony services in the Southeast.

The service initially provided movie reviews, a content area devoid of much of the controversy surrounding the issue of defining news. Nevertheless, the relationship between the newspaper company and BellSouth soon became contentious. The newspaper company charged BellSouth with failure to live up to its promotional timetable and said BellSouth was unwilling to share pertinent market information. BellSouth denied the allegations, but the incident is an example of how distrustful the newspaper industry was of the telecommunications industry at this time.

David Easterly, president of Cox Newspapers, testified before a congressional subcommittee that his experience convinced him that if allowed to enter the information services business directly, the telephone companies would use the control of the network to create "a home field advantage." He added, "To hurt other players [regional Bell companies] can be just a little bit slow in handling service problems. They can drag their feet in sharing market data. In subtle ways, they can deploy their own advanced technology to favor their own services."[62]

The notion that the telephone companies would use their own technology to their advantage, while creating a disadvantage for competitors, was a recurring theme of the newspaper industry argument. However, it was usually theoretical posturing. The allegations were taken more seriously when the president of a large newspaper company testified that his company actually had experienced heavy-handed tactics by one of the Bell operating companies.

Action Versus Inaction

Judge Greene in 1987 reaffirmed his ruling that the "diversity principle" was an important construct to protect, and continued to restrict the information services business in which AT&T and its offspring could engage.[63] The ruling angered AT&T and Bell officials who "had launched a major campaign to be freed from the restrictions in the consent decree," but newspaper executives again declared victory.[64] This time, an industry executive from within the trade association's leadership argued the victory came with an obligation to invest in developing new services. Johnson wrote that "newspapers must take action or forever be prepared to live with the consequences of inaction." He further explained his position:

> . . . both the court and perhaps Congress will also be looking to see if the pro-
> ponents of the Diversity Principle are actually willing to invest in development
> of the information industry, or whether they are simply using policy arguments
> to protect their own vested interest. . . . The time for standing behind a policy
> position is over. Newspapers will be asked to put up or shut up.[65]

Johnson's remarks may have reflected his personal passion, but he did not provide a convincing portrait of an industry about to be mobilized into action. Instead, the newspaper industry during this period drifted toward ambivalence regarding its role in the electronic services marketplace.

Johnson described the mindset of many within the industry as "dangerous, myopic and not in touch with what is truly happening in the marketplace." He observed that court rulings favoring the newspaper industry's position against AT&T had fostered complacency within the newspaper industry. When combined with the decisions of Knight-Ridder and Times Mirror to close their video-text projects, such complacency, Johnson said, resulted in a false sense of reality. He wrote that some "take comfort in the fact that electronic publishing seems to be a technology in search of a market. They conclude that there really was no threat to or opportunity for newspapers in the first place, making all the discussion surrounding Greene's decision so much wasted energy."[66] Johnson's activist rhetoric did little to sway the thinking of most leaders within the newspaper industry. Instead, industry leaders at the close of the 1980s positioned the newspaper industry's competitive circumstances in much less dire terms.

An industry conference held in Chicago in 1989 provides an example. Rather than issuing a call for action, Blethen told attendees that threats to newspapers from electronic information services were long term in nature. "Technology won't replace us. Our readers aren't about to start calling up electronic newspapers on their computer screens," he said. "There is no single competitor that will crush us and no single bold stroke that can protect us. . . . To respond, every newspaper will have to take multiple steps. Solutions will vary from market to market."[67]

In a later presentation, Blethen deployed a phrase that summarized the newspaper industry's newfound approach to its telecommunications policy dilemma: "The threat is more imminent than the opportunity."[68] The phrase signaled that newspaper companies needed to continue to support the association's lobbying effort to keep the telecommunications threat in check during the short term, but that it was also understood that newspaper companies were not expected to invest too heavily in technology until profitable markets emerged.

Therefore, as the 1980s came to a close, the newspaper industry was in a holding pattern in regards to its next technology initiatives. Instead, attention shifted to fighting a new enemy. As the telecommunications industry evolved, the threat from AT&T diminished. But the size and scope of the regional Bell operating companies increased significantly, which shifted the focus of the debate. The following section explores the rise of the Bell companies and how the newspaper industry responded to a scenario it had not envisioned when AT&T was dismantled.

The Bell Uprising

The end of the 1980s brought little in the way of resolution in the ongoing policy disputes between newspaper companies and the nation's telecommunications industry. If anything, the war of rhetoric grew more intense as the focus shifted away from AT&T in favor of the regional Bell operating companies that were established in the aftermath of the AT&T divestiture in 1984. In 1989, Judge Greene issued new rulings that kept these regional Bell companies from directly participating in information services. He stated, "There cannot be the slightest doubt that, should the regional companies be permitted to engage in information services on a more substantial scale, they would in short order dominate the information services market."[69] AT&T had filed its own petition to be freed from the restrictions,[70] and eventually Greene relented in the case of AT&T, allowing that company into the electronic publishing arena. AT&T had already begun participation in a consortium along with Time Inc., Chemical Bank, and Bank of America on a proposed service built around home banking called Covidea.[71] There also were mounting efforts in Congress to back legislative reform that would overturn at least some of Greene's restrictions on the regional Bell companies.[72]

Once again, the newspaper industry's trade association prepared its members for an intense legislative fight on Capitol Hill, framing the industry's position as a defense of the "diversity principle" that requires "a separation of content and conduit."[73] The regional Bell companies decided to take the fight directly to the newspaper industry, filing briefs claiming that the court's refusal to allow them into the information services business violated their First Amendment rights.[74] An author specializing in the communications industry defined the two positions:

> The [regional Bell companies] say they're being denied their First Amendment rights and should be allowed the freedom to publish electronically or otherwise as they see fit. [The newspaper industry] argues that the First Amendment guarantees a diversity of free expression and that diversity would be subverted by allowing the [regional Bell companies] to enter the field as the provider of both the content and conduit of electronic information.[75]

It was essentially the same argument that had prevailed against AT&T earlier in the decade, but the evolution of the marketplace and advances in telecommunications technology caused judges to look at the situation differently.

In 1990, the U.S. Court of Appeals said Judge Greene was wrong in keeping the regional Bell companies out of the information services market and sent the issue back to his court for a retrial. Following nearly a year of additional legal wrangling, the regional Bell companies were free to enter the information services business using their own transmission lines in their own markets.[76] This turn of events triggered another intense lobbying skirmish between the newspaper industry and the regional Bell companies as the newspaper industry once again sought legislative intervention.

The newspaper industry returned to its earlier tactics and attempted to portray its opponent as the telecommunications equivalent of Goliath. Entering 1990, the seven regional Bell companies combined for more than $77 billion in revenue, which included at least $6 billion in Yellow Pages advertising. The nation's 1,600-plus daily newspapers had combined revenue of about $45 billion.[77] Positioning itself as an industry that could be crushed by a giant, the newspaper industry launched an advertising campaign in many of the nation's leading newspapers designed to sway public opinion. The newspaper industry was joined by other businesses and trade associations with similar interests, such as the NCTA and Dialog Information Services, in a print campaign that featured the slogan: "Don't Baby the Bells. Keep Competition Alive."[78] The ads read in part:

> If the regional Bell telephone companies are permitted to provide their own information services, they will have the ability—and the incentive—to compete unfairly with the companies they now serve. The Bells will deny competitors the latest technological advances. They will even find ways to make telephone ratepayers foot the information-services bill.[79]

The regional Bell companies countered with their own ad campaign aimed squarely at the newspaper industry with the theme: "America's Future. Too Important To Leave On Hold." The campaign portrayed the newspaper industry as an obstacle to progress, claiming that history revealed how newspapers tried to stop radio, then television, "and now they're trying to stop the benefits of the information age."[80] The Bell ads read in part:

> A revolutionary array of information services could be available to the American public through the regional Bell telephone companies. Americans could enjoy broad and affordable access to crucial information services in the worlds of education, medical services and entertainment. Many of these benefits are already available overseas. Yet America's largest newspapers are fighting to deny them to the American people. Why? Because they fear the threat of competition. They are reacting as they historically have when new technologies offer people new information choices—radio, cable television and now, even new uses of the telephone.[81]

Despite such posturing and the introduction of several proposed bills, neither side could convince Congress to act in the early 1990s. This was largely due to the rapidly changing telecommunications landscape, which saw the emergence of new long-distance carriers, local market telephone competition, data-specific networks, and a rapidly expanding cellular industry. "Technology has superseded rules governing the industry broadly and the telephone companies specifically," said a government affairs executive with US West, then one of the regional Bell companies.[82] By the early 1990s, the rapid changes in the telecommunications industry made the argument between regional Bell companies and the newspaper industry over "information services" look almost archaic if not irrelevant.

Lawmakers and regulators struggled with how to regulate the new telecommunications marketplace, so much so that it would take Congress until 1996 to rewrite the nation's principal telecommunications law. And by then the Internet had begun to exert its influence on the market.

Discord Replaced with Partnerships

As the '90s began and the newspaper industry took stock of its electronic publishing efforts, there were few tangible results following a decade that included periods of experimentation and investment and a relationship with the telecommunications industry that alternated between outright enmity and wary partnerships. James Lessersohn, manager of corporate planning for the New York Times Co., described the assortment of electronic publishing projects underway in the newspaper industry at this time as "augmentation media," which included low-cost videotext, voice services, and news summaries delivered by facsimile.[83]

Cabletext had fallen out of the mix. And in regard to videotext, Lessersohn stated that it represented a medium "a lot of people in our industry are trying to forget."[84] Facsimile editions had limited consumer appeal because most fax machines were located in businesses, which discouraged personal use such as receiving a "faxpaper." Perhaps there was also a bias against a "new" medium that was not new at all, but one recycled from the pre–World War II era. As Light noted in a history of the facsimile, "from the 1920s through the 1940s . . . facsimile was one of the most exciting innovations of its day,"[85] even attracting the participation of some of the nation's leading newspapers including the *New York Times*.[86] But more ironic than the return of facsimile-based newspaper editions was that much of the development of this "augmentation media" was undertaken through partnerships between newspaper companies and telephone companies.

At first, there was the perception that such arrangements were tantamount to breaking from the ranks and joining the enemy camp. For example, the publisher of the *St. Louis Post-Dispatch* said his company's discussions about projects with one of the regional Bell companies upset some of his industry colleagues. "We wanted to be loyal members of the club, but our own business interests had to come first," he said.[87] An executive with BellSouth, one of the regional operating companies, said other publishers expressed similar sentiments. "A lot of them believe there's a way that we can work together," he said, but with the lobbying battle that had taken place, "they don't want to be the first or maybe even the second."[88] By late 1993, however, newspaper companies were largely over such concerns. Newspapers and the regional Bell companies had more than a dozen joint projects underway.

For example, the *Chicago Sun-Times* partnered with Ameritech to create a phone-based fantasy baseball service; the *Rocky Mountain News* in Denver created a health newsletter delivered via fax in cooperation with U.S. West; *Newsday* teamed up with Nynex to launch an audio news service that used voice mail; and a group of newspapers in Utah formed a consortium to create a classified advertising network with U.S. West.[89] Cox Newspapers, through the *Atlanta Journal-Constitution*, teamed up with BellSouth to offer voice information through a 511 service, a direct three-digit dial number intended to boost usage of audio services by providing an easier number for consumers to remember.[90]

Dow Jones & Co., the publisher of the *Wall Street Journal*, was perhaps the most prolific dealmaker with phone companies, even creating a "Telco Alliance" department.[91] It had projects with at least four of the regional Bell companies, including a reader line service in partnership with BellSouth and a mobile information service with Southwestern Bell.[92] The Dow Jones manager of those alliances said the attitude regarding newspapers and telecommunications had evolved due to market realities: "The industry has matured. We have a better understanding now of what is involved with information services, and we're better able to protect ourselves. The phone companies have gone through a similar maturation process. They realize they have to work with others."[93] A telecommunications executive concurred: "as companies explore new technologies, former adversaries sit down together and find it's in their interests to make a deal."[94] Newspaper companies had pursued their videotext projects of the early 1980s independently, while the industry's short-lived incursion into the cable television business was largely through partnership deals, as discussed in the previous chapter. The partnership model was carried forward in the latter half of the decade when newspapers set aside their differences with the telephone companies to bring several information services projects to market. By this juncture in the evolution of online media, newspaper companies had settled on a strategy best described as one of risk management. The industry wanted to participate in online information delivery, but it wanted to mitigate the financial investment and level of effort allocated to such ventures.

A survey of 250 newspaper industry executives in 1992 found that 61 percent of the respondents felt that newspaper companies should enter joint ventures with telephone companies as a way to expand into electronic information services. Perhaps the more telling result, however, was that only 20 percent of respondents said they actually planned any such venture with a telephone company.[95] The

ANPA chairman wrote in January 1992 that the association's lobbying effort would continue as a way to protect the industry's interests, but the focus of the rhetoric had shifted. He wrote: "The point is not to bar the Bells from information services forever. . . ."[96] And in an act that was perhaps most illustrative of the changed environment, AT&T joined with the newspaper industry that year in signing a "unity statement" encouraging Congress to enact legislation that reflected the spirit of consent decree that had broken up the telephone monopoly.[97]

Bogart believed that the negative experiences of the newspaper industry's own projects during the 1980s caused many executives to question the rhetoric regarding telephone companies as eventual competitors. He wrote that many publishers wondered: "If these all flopped, why should the phone companies have better luck?"[98] However, if the telephone companies were to generate any market traction, newspaper executives figured that—at least in the short term—it would be due to the fact that telephones were the entry point for online services. As personal computer adoption expanded, so did the adoption of modems used to connect home telephone lines to computer-based services. It was a technology trend the newspaper industry could not ignore. As the ANPA chairman stated at an industry technology meeting: "Telephones are tomorrow's new medium. Wherever you look in the next few years . . . we're going to see new uses of telephones."[99] The recognition of this market condition led the newspaper industry to soften its hardline opposition to the telecommunications industry.

Given that the telephone technology had assumed this key role, the public mostly viewed the telephone industry as progressive, while the newspaper industry was perceived as technically deficient. Industry leaders began to address the negative perception proactively. During an industry address, one key leader stated bluntly that "if we become defensive and rigid in dealing with change, we will wither," adding that newspaper companies "cannot direct all of our energy and resources into attempts to force the market to accept a medium that in some cases is simply not the best for its needs."[100] He explained, for example, that printed newspapers could not serve a businessperson's need for real-time stock information.

It was such tacit admissions of weakness that led to a resurgence of activist rhetoric. Easterly, the president of Cox Newspapers, warned his industry colleagues: "Newspapers that have not been investigating electronic avenues of information—and that's most of them—had better get on the ball now."[101] Nevertheless, the technologies mostly under consideration by newspapers at this time were described earlier as augmentation media such as fax services and inexpensive bulletin board systems. An industry executive explained that these technologies largely were considered "temporary platforms," or transitional technologies that would help "prepare us for that new market," but he admitted that few in the newspaper industry could envision what that new market would entail.[102]

Discussion

Newspaper industry executives arrived in the early 1990s with an assortment of emotions. There was frustration that a decade-plus of experimentation and hundreds of millions of dollars in collective capital spending had failed to produce any significant breakthroughs in the electronic information arena. Nevertheless, a sense of relief also permeated the industry's executive suites. Printed newspapers had survived—some would argue thrived—in the 1980s, a decade that was supposed to have ushered in an electronics-based Information Age. The ongoing financial success of the core printed product contributed to the disdain many newspaper executives had for projects such as electronic bulletin boards and facsimile editions that attracted users only in the hundreds and low thousands—far from the mass media numbers they were accustomed to selling to advertisers.

As the telecommunications landscape continued to evolve, the newspaper industry softened its lobbying stance against the telecommunications industry because it no longer considered the telephone companies to be as scary as they were previously perceived. Part of this change of attitude was due to recognizing the role telephone technology was playing as the entry point for online services. Newspaper executives also understood that their lobbying war with the telephone industry throughout the 1980s and early 1990s had damaged the newspaper industry's public reputation. The telephone industry's rhetoric that positioned the newspaper industry as an obstacle to progress resonated far more deeply with the American public than did the newspaper industry's characterization of the telephone companies as out-of-control monopolies.

The newspaper industry's attitude shift forms the backdrop for the next chapter, which explores another attempt by the newspaper industry to exploit online media. Many in the newspaper industry were hungry for a breakthrough in electronic information services that had so far eluded them. Therefore, when proprietary online systems began to accumulate a critical mass of subscribers, newspaper companies became very interested. They wanted to learn if a viable electronic distribution model had emerged.

Newspapers still were not interested in making huge financial investments in their own electronic services infrastructure, which made partnerships with the proprietary online companies an attractive alternative. Furthermore, publishers recognized a business model that made sense to them: centrally controlled content partially subsidized by subscribers and supported by advertisers. Rather than turn adversarial as it had with the telephone companies, the U.S. newspaper industry almost seemed in a rush to embrace these systems, even though some had the backing of giants such as IBM. These services, including Prodigy, AOL, and a revamped CompuServe, appealed to newspapers by achieving momentum in the marketplace with a value proposition to consumers that eclipsed the earlier failed videotext projects. The newspaper industry's relationship with these companies is the subject of Chapter 4.

4

Newspapers Embrace Proprietary Online Services

U.S. newspaper companies invested more than $100 million in online newspaper projects in the early 1980s,[1] but the most ambitious efforts were closed by the middle of the decade. They had failed to generate a sufficient audience to sustain them. In subsequent years, newspaper companies deployed a disparate collection of experiments ranging from audiotex and fax services to online bulletin boards as the industry struggled to find its role in the electronic marketplace. Furthermore, the industry's political lobbying against the telecommunications industry's expansion into the electronic information services left the public with the impression that newspaper companies were obstructionists rather than innovators.

Against this backdrop, this chapter explores a relatively brief period in newspaper history when newspaper publishers turned to a new group of partners they believed would help them capitalize on electronic information opportunities that had so far proved elusive. These companies became known collectively as proprietary online services, and they sought to create an online mass consumer market where others had failed. Researchers said newspapers pursued partnerships with companies such as Prodigy and AOL because they were seen as mutually beneficial at a time of nascent market development:

> The Catch 22 of early electronic newspaper services was that consumers did not want to subscribe if a wide range of services was not available, but information

providers did not want to spend development money on systems that had no subscribers. The linking of electronic newspapers with commercial information services bypasses this problem.[2]

Still, the decisions by newspaper companies to partner with these companies were complex, due at least in part to the uncertainties of technology direction and concerns over the financial stability of these newcomers.

This chapter examines the newspaper industry as it navigated through an array of confusing choices during this period. A discussion of market and industry conditions provides context for the industry's response and frames the exploration of business concerns that guided the complex decision-making process. The chapter principally deals with the relationships that emerged between newspaper companies and proprietary online services. The chapter discusses several companies that were active partners with the newspaper industry during this period, with particular attention given to AOL and Prodigy. These two proprietary online services emerged as the most influential in their dealings with the newspaper industry.

By examining the newspaper industry's rationale for partnering with proprietary online services rather than creating such services on its own, the industry's conservative business culture is further revealed. Shifting attitudes regarding technology helped newspaper executives overcome the stigma of the failed Viewtron project, but not to the extent that they were willing to venture into the online media business on their own. In partnering with proprietary online services, newspaper companies found a comfortable alternative that allowed them to provide content while others worried about the technical infrastructure and the investment it required. Finally, this chapter serves as a bridge within the overall book by focusing on a period that connects the newspaper industry's videotext past and its Internet future. Newspaper executives had turned away from their protectionist rhetoric that had dominated the late 1980s and again were venturing into the online arena slowly and on their own terms. But their reliance on the new partnerships would be brief, as the emergence of the Internet forced newspaper companies to confront a new marketplace once again.

An Emerging Market

When Knight-Ridder launched its Viewtron project in the early 1980s, the notion of a computer-based information service was still in the realm of science fiction to most Americans. Not only did Knight-Ridder have to develop and market the distribution technology, it also had to explain what exactly its service was for and what it did. But during the decade following the Viewtron failure, consumer exposure to computers had increased appreciably. The *Washington Post* wrote that "Americans are acquiring skills that the new services . . . require," noting that "more and more office workers and students use PCs; tens of millions of Americans have home video game units . . ., which make them comfortable interacting with a screen."[3] More importantly, the increased exposure to computers affected consumer attitudes regarding electronic information services, which the *Post* article reflected: "Americans' attitudes toward information are changing in ways that will create demand. They are comfortable getting it off a screen. They want it in greater quantities and variety, delivered faster, at lower cost."[4] Newspaper publishers had invested heavily in new technologies to operate their own companies, so they could see firsthand how computers in the workplace influenced people's general attitudes about technology. Publishers were also aware that the number of home computer users had grown beyond the hobbyists, which in turn provided a promising mass audience for online information services.

The Videotex Industry Association reported that by the late 1980s about 700,000 homes were connected to at least one of the 40 "fee-based consumer videotex systems" known to be operating. The group said 500,000 homes also were using free electronic bulletin board systems. After accounting for crossover users between the fee-based and free services, the trade association concluded that 960,000 households were tapping into information via online services. The market for these services was bolstered by successful efforts such as the WELL to attract leading-edge thinkers and cultural observers. Created in 1985 as the Whole Earth 'Lectronic Link, this service became symbolic of what Rheingold described as "virtual communities."[5] He wrote:

> Like others who fell into the WELL, I soon discovered that I was audience, performer, and scriptwriter, along with my companions, in an ongoing improvisation. A full-scale subculture was growing on the other side of my telephone jack, and they invited me to help create something new.[6]

The user base of such services accounted for only about 1 percent of all households in the U.S. by the end of the 1980s, but the growth trend was encouraging, fueled in part by the sense of wonderment that Rheingold expressed. The trade group forecasted that by 2000, 97 percent of the country's households would be connected to such online services.[7]

Such lofty prognostications led to a resurgence of the optimistic rhetoric surrounding the market for online information services. This time, however, the rhetoric regarding the prospects for an online information marketplace in the U.S. was tempered somewhat by the experiences of the early 1980s, as illustrated by this example:

> Gingerly referring to the past failures of videotex in this country, [proponents of consumer videotext] say the industry has matured and learned valuable lessons about marketing and distribution. And they note that once skeptical lawmakers and regulators have begun to see the need for videotex if the U.S. wishes to compete in the 21st century.[8]

The last point of this statement regarding U.S. competitiveness should not be overlooked when discussing the rhetoric surrounding the development of the country's online services market. The failure of several early consumer-focused projects had stalled momentum in the U.S., causing some to question the country's commitment to develop an electronic information market. Kinsley, for example, suggested that the U.S. was in danger of losing its "international competitiveness" because the market for online services had been slow to develop in comparison to Western Europe, especially France.[9] Others argued that the market in Europe was not an accurate comparison because state-owned telephone monopolies there had subsidized online information services and provided consumers with incentives to use the systems.

Efforts to import and deploy technology from some of Europe's successful systems such as France's Minitel met with specific resistance in the U.S. market. For example, an executive from a U.S. telecommunications company said that borrowing business models or technology from elsewhere "is to have failed in the effort to be the world leader in technology."[10] In the face of such polarizing rhetoric, the emergence of U.S.-based services such as Prodigy and AOL was important. These companies diffused the discussion about American technology, clearing away another obstacle that some believed had contributed to slow market development.

Information Today, an industry trade publication, declared 1989 as the year when a concerted effort began to make online services "attractive to consumers."[11] And in the early 1990s—the years immediately preceding the Internet's emergence as a consumer platform—Prodigy, AOL, CompuServe, and a few smaller players represented the promise of the Information Society. Consumers could use home computers to connect affordably to a vast array of information databases and electronic services just as the pundits had predicted. AOL's slogan "You've Got Mail" became symbolic of its time.[12] Collectively, these services attracted more than 3.5 million subscribers during this period.[13] Kyrish reported that by 1993—five years after its official commercial launch—Prodigy had attracted about one million subscribers, while two other services—CompuServe and AOL—had 1.5 million and 285,000 subscribers, respectively.[14] Kyrish wrote that "an historical review of the period between 1989 and 1993 suggests steady but hardly explosive growth."[15] Nevertheless, the growth created an audience significant enough to cause newspaper publishers to take notice.

The emergence of these proprietary online services coincided with a period in the newspaper industry when publishers were especially introspective. Newspaper companies struggled during the economic recession of the early 1990s, but publishers were considering their next moves as business conditions showed signs of improving. As the newspaper industry's interest in the electronic marketplace began to ramp up in the early 1990s, the *American Journalism Review* attempted to place the activities in a proper societal context:

> Certainly, there's a bit of the "millennium syndrome" afoot, as the approaching turn of the century makes people feel they're on the brink of a new age, with a Task Force 2000 forming in almost every industry. The problem is that no one really knows how newspapers will be read and distributed in the next 10 or 15 years, or even five years. . . . But the industry is in hot pursuit of the answer.[16]

An industry executive said the opportunities presented by the propriety online systems created "exhilaration and enthusiasm" in the newspaper industry, adding that advertising prospects were favorable because "everyone has had their consciousness raised about the information highway."[17] The following section examines the status of the newspaper industry during this period and discusses the factors that led many newspaper companies to partner with proprietary online systems.

Newspapers in the Early 1990s

Even though profit margins in the newspaper industry remained strong relative to other industries, many issues had publishers rethinking their approach to the business. Declining circulation and eroding market share, combined with the lingering effects of economic recession that had depressed advertising revenue, forced publishers to consider that their industry was in the midst of fundamental change.

Underwood observed that "amid all the flailing about as newspapers prepare for an uncertain tomorrow, three general strategies" emerged: "effort[s] to save the newspaper as it is, efforts to augment the newspaper electronically, and efforts to look beyond the newspaper-on-print."[18] Several examples illustrate his first category, such as shorter stories, more graphics, and increased use of color photography. He also cited Gannett Inc.'s "News 2000" program, which emphasized coverage of community issues, as an example of the efforts aimed at redefining content to improve readership.[19] The "efforts to look beyond" print were deemed too futuristic and were pursued half-heartedly by the industry. Underwood discussed Knight-Ridder's investment in researching and developing the prototype of a flat-panel electronic tablet with a touchscreen as an ambitious illustration of innovative thinking. A contemporary media commentator recalled that project as "an eerily prescient 1994 vision of" Apple's iPad released in 2010.[20] At the time, however, Knight-Ridder's tablet concept was stymied by the technical shortcomings of its era and an industry culture that viewed such projects as science fiction. Most newspaper executives could not envision a future without the printed newspaper form at the forefront. Therefore, most activities undertaken by the newspaper industry during this period took place within Underwood's second category: "efforts to augment the newspaper electronically."[21]

Within this category, newspapers began to think seriously again about an electronic future and how those activities could be integrated into existing operations. However, the desire to find business models that would prevent the financial disasters associated with the failed online efforts of a decade earlier was paramount. Despite the newspaper industry's financial profitability during this period and the bravado it exhibited during the political battle with the telecommunications industry, newspaper publishers were scarred by the memory of financial losses from those early online projects. This aftermath influenced the decisions regarding how newspaper companies would proceed with electronic information services.

Viewtron's Lingering Effects

Newspaper companies—wary of placing too much emphasis on the electronic information marketplace—preferred to take small steps rather than embark on bold moves. The Viewtron failure loomed large: as the *New York Times* put it, "[Viewtron] has been cited ever since by skeptical news executives as a warning that electronic ventures can be business disasters."[22] In monetary terms, the $50 million that Knight-Ridder had invested in Viewtron for no return had little material effect on the company's financial position. However, in the conservatively managed newspaper industry, any financial loss was difficult to accept, and over time, the Viewtron project became remembered not as a pioneering effort, but as a financial boondoggle.

"The scars from Viewtron are still very vivid," said an editor at a Knight-Ridder newspaper. "It took a while to get it into people's minds that we should get back into electronic distribution."[23] Another industry executive echoed those remarks: "That project poisoned the water for all of us. All people could see was the red ink."[24] With negative sentiment regarding the Viewtron experience so pervasive in the industry, proponents for new initiatives had to find ways to promote the potential for success rather than allow the industry to continue to dwell on past failures.

Some have suggested that the modest financial success of low-cost bulletin board services such as StarText operated by the *Fort Worth Star-Telegram* helped the newspaper industry to once again think about electronic distribution in broader terms. "Just because we built a few Edsels doesn't mean the car is wrong," said the StarText marketing director.[25] By demonstrating that not all online projects were money losers, these bulletin board systems represented the potential for more robust systems to be developed within a profitable cost structure. As the industry began to shake off the effects of the early 1990s economic recession, newspaper companies seriously began to explore again their options for returning to the electronic arena in a more meaningful way. The next section will show how these activities represented a significant shift in the newspaper industry's approach to the online services market.

Shifting Newspaper Industry Attitudes

The newspaper industry's reluctance to invest directly in electronic information services seemed to be even more pronounced during the most heated periods of the industry's political battle with the telecommunications industry. The lingering effects of Viewtron may have been a contributing factor, but newspaper companies had evolved into defensive operators rather than strategic planners. As one industry executive stated:

> For a while there, newspapers were primarily identified with blocking the Baby Bells. It gradually dawned on newspapers that they couldn't—and shouldn't—depend on Judge Greene to save them from the future. That defensive kind of strategy was ridiculous. We should recognize these changes and go on.[26]

This statement underscored the shift in the industry's approach. It represented a new way of thinking and marked a break with the latter half of the 1980s when the industry faced a lack of direction regarding the future of electronic information. Industry leaders were once again seeking an active role for newspapers in the development of the electronic information market.

Researchers had suggested that the industry's historical approach had less to do with the emotions of its leaders and more do with the process-driven nature of the industry. An industry manufacturing and distributing a new product every day was by its nature internally focused. It had little intrinsic interest in long-term strategic planning.[27] To alter the industry's dynamics, Wilson and Igawa wrote that newspaper companies needed to embrace systemic change if they expected to succeed with new information services. They concluded:

> The hallmark of innovation and new ideas is ambiguity, asking upside down, inside-out questions, then shaping ideas in tandem with the people you want to serve. Overall, the rigid routine of a newspaper works against that process. . . . So newspapers adapt pragmatically, feeling their way, which accounts for the piecemeal response to the relentless tides of change.[28]

The industry's structure was also a factor contributing to the "piecemeal response." The term "newspaper industry" is used throughout this book, but as noted in the introduction, its usage refers to a collective of newspapers that vary in size, format, financial strength, and corporate ownership. These newspaper companies rarely competed directly with each other in a local market, but their variety also made cooperation difficult to achieve. When newspaper companies did rally around a common technology initiative, it was recognized as another sign of the industry's shifting attitude.

In May 1993, 17 companies, including Gannett Inc., Knight-Ridder Inc., Tribune Co., Hearst Newspapers, and Times Mirror Co. founded a consortium called the "News in the Future" project at the Media Laboratory of the Massachusetts Institute of Technology (MIT). Designed as a five-year project, the consortium planned to spend up to $2 million each year researching emerging electronic technology for delivering news to consumers. Several interested parties outside the newspaper industry also joined the project, including computer manufacturer IBM, broadcaster Capital Cities/ABC, regional telephone provider BellSouth, and advertising agency McCann-Erikson.[29] The participation of an advertising agency was seen as especially significant given that a goal of the project was to develop ways for news and advertising to form "a seamless information landscape that ranges from the most serendipitous to the most urgent information."[30]

An industry executive involved with the consortium said it was needed to bring the industry together and address technology development in a coordinated way. "There are a lot of ants running off in different directions. But I [now] see a different anthill being built . . . we all were going down separate streets . . . now it's all sort of coming together and we need to sit down together to operate it properly."[31] Boczkowski wrote that a consortium of newspaper companies collectively investing in research and development represented a fundamental shift in behavior because it is "an industry not used to investing money in this type of activity."[32] He quoted an industry executive who noted how exceptional the MIT project was: "I don't think anything like this has ever happened in the newspaper business before."[33] This activity represented a tangible example of the changes Wilson and Igawa had suggested was necessary for the industry to succeed in electronic services.

Nevertheless, the ongoing struggle between the industry's recognized need for innovation and its deeply rooted conservative business practices was evident as the events unfolded in the early 1990s. The following section examines the business discussions taking place during this period and how they pointed to partnerships with proprietary online services as the only logical conclusion to the collective thinking during this period.

Business Concerns

The newspaper industry's newfound approach—proactively researching the possibilities of electronic delivery—met with positive reactions from the financial and investment communities. The prevailing sentiment on Wall Street was that emerging electronic competitors could undermine the newspaper industry's advertising pipeline, especially its long-established dominance in classified advertising. As one industry analyst stated: "The threat of these developments is reason enough for newspapers to invest in electronic publishing, especially now when they have the money to do it."[34] In fact, many newspaper companies—especially those that were publicly traded businesses—felt pressured by investors to shore up their defenses against a perceived onslaught of new electronic competition.

But how to accomplish that while also delivering the relatively high profit margins investors had come to expect from the industry left many executives exasperated. The comment from a senior executive with the New York Times Co. illustrated the frustration: "We've got the media on our backs, Wall Street on our backs. . . . But I don't know how to spend $1 billion [on electronic services] and make it pay out."[35] This comment also suggests that even though newspaper executives in key positions at the major U.S. dailies were not immune to criticism, they also were not interested in investing heavily in new technology simply to silence their critics.

Newspaper publishers may have recognized the potential of a threat from new electronic competitors, but they did not consider the threat to be imminent. Spending too heavily—and possibly spending on the wrong technology—was a risk they were unwilling to take. One comment in particular summarizes how afraid the industry was of making the wrong investment: "If you can show me somebody who says, 'Yeah, I know exactly what the future holds,' there's a good chance he won't be in business soon."[36] The comment indicated how little interest there was for pioneering. Instead, the newspaper industry remained committed first and foremost to its printed format.

Even an executive charged with managing electronic information projects at a major daily expressed the prevailing opinion: "There's nothing wrong with a mature business, if it's managed properly. We in the newspaper business shouldn't give up our day jobs."[37] These sentiments reflect Boczkowski's observation that although newspaper companies wanted to participate in the electronic marketplace—and even saw a competitive need to do so—the "pragmatic" approach took precedent. He wrote: "American dailies were often more interested in the short-term health of the core print business than, more idealistically, in projects that seemed more promising with comparatively higher payoffs that could only pan out in a longer term."[38] Financial analysts, industry commentators, and academic researchers realized that despite the creation of a research consortium and an increase in rhetoric about exploiting electronic information services, the U.S. newspaper industry had no appetite for grand innovation on its own.

The majority of newspaper companies seemed most comfortable with a partnering approach that limited capital investment, but positioned them to reconsider if market conditions changed. An executive with the *Los Angeles Times* described the strategy: "If there's a big upside for newspapers on the so-called superhighways of the future, we intend to participate in it. If there's a downside, we should, with these [partnering] programs, be able to cushion it."[39] Therefore, as the newspaper industry expanded in the area of electronic delivery, it did so in a way that mitigated risk. Boczkowski observed that "the bulk of activity in the period 1992–1994" involving electronic or online newspapers "took place in relation to online services."[40] Most newspaper publishers liked an arrangement where the online companies would manage the technology infrastructure, while the newspaper companies would provide content and share in the advertising revenue. The following section explores how this partnering strategy took shape.

Partnership Strategy

The decisions to pursue partnerships with the proprietary online systems were not without controversy, and there was considerable tension within the newspaper industry regarding the best course of action to take. Newspaper companies were noted for their independence, and moving forward with a strategy based on partnerships required a level of cooperation to which most of these companies were unaccustomed. One executive observed: "Those of us in a business in which we own the printing press have a certain level of discomfort about using somebody else's."[41] The realization that newspaper companies would have to give up—or at least share—control led to concerns about how to structure the partnerships. For example, an executive advised his colleagues to enter such arrangements only if they understood that "issues of control and management can be so complex and divisive that the ties holding non-traditional partners together can snap if the match isn't right."[42] The newspaper industry had been down this path before, however. The arrangements with cable television companies in the early 1980s, for example, involved partnerships that were just as complex, if not more so, than the deals contemplated with the proprietary online services.

Nevertheless, the harshest criticism came from those who believed that the industry's direction was strategically short-sighted. For example, one outspoken critic charged that partnering with the proprietary online services was tantamount to "hauling your presses off to the dump and hiring your competition to print your paper," adding that "the only way to protect ourselves from more of an invasion by electronic services is to start those services ourselves."[43] But such opinions were in the minority and did not deter the industry's direction. There were no companies within the newspaper industry at this time willing to invest the resources to either acquire a proprietary online service or to create one.

In most cases, partnering with the online proprietary services was viewed as the most prudent option available. One executive with the Tribune Co. said his company had decided that a partnership would allow it to think differently about the role it plays: "Our role as a company wasn't to develop technology but to take technology that others had developed and make it useful."[44] Newspaper companies strategically justified their partnership decisions with this line of thinking—one that separated concerns about production, distribution, and content.

Most of the deals that were signed during this period followed a similar pattern. The deals allowed newspaper companies to control the content they provided, while sharing production responsibilities. Distribution largely was the purview of the proprietary online services. The biggest issue, however, involved how the newspaper companies would be compensated for their content, which is explored in the following section.

Deal Terms

When asked by an industry trade publication why he had agreed to partner with one of the online services, the president of Cox Newspapers said it was simple: "Greed. Because we're going to make some nice money on this."[45] But newspaper executives understood that making money on these endeavors required striking the proper balance between the revenue they would receive and the investment they would be required to make in gathering and producing the content. A sales executive for the *Hartford Courant* described the conundrum confronting newspapers this way: "How much to invest and what return to expect have never been more fuzzy. Newspapers aren't going to receive the returns to which we are accustomed, and that makes everybody nervous."[46] Newspaper companies understood that partnerships designed to mitigate risk also meant some limit to the upside of the market, but the challenge was to get the most favorable terms possible. Industry analysts, however, were concerned that some of the earliest negotiations favored the online services rather than newspapers.

Analysts feared that the newspaper companies were not doing enough to protect their franchises should the proprietary systems succeed as local information providers, supplanting newspapers in their own markets. "Don't let on-line services cannibalize your readers," one analyst warned.[47] A business development executive at Gannett's *USA Today* framed the question facing the industry this way: "Who owns the customer?" She added: "Newspaper publishers don't want their products to be thought of as mere mastheads in someone else's mall."[48] At the very least, industry analysts wanted the newspaper companies to be more aggressive in negotiating the split in subscription revenue. An analyst told publishers they should increase their share of subscription revenue from 10 to 20 percent, common in the early deals, to at least 50 percent, arguing that newspaper companies involved in the early deals "are getting undercompensated" for bringing new users to the electronic services.[49] Subscriptions, however, were considered an ancillary source of revenue in the newspaper business, and this was no different when publishers negotiated their online deals. Advertising was viewed as the most important revenue stream, and executives wanted to get the best deal for this aspect of their online partnerships.

One industry executive urged his colleagues to accept nothing less than a 50 percent split of the advertising revenue in their online partnerships. He argued: "Advertising is one of our biggest advantages," adding, "Don't make a bad business deal."[50] The newspaper industry's efforts to negotiate favorable terms and avoid the bad deal were exacerbated by the selection of possible partners.

Companies such as AOL, Prodigy, and CompuServe are remembered because they achieved the largest audiences during the early 1990s. But newspaper companies had numerous other potential partners to sort through, including such companies as Delphi, Interchange Network, and General Electric's GEnie. The following material examines the choices available to newspaper companies and how publishers went about selecting an online service partner.

Confusing Choices

The newspaper industry's official trade publication observed that in the ten years following the first wave of videotext projects, online options available to newspaper companies had "exploded," adding that "commercial online services are proliferating and starting to compete for newspaper content."[51] As newspaper companies began to reengage with the marketplace for online services, many executives found the assortment of options bewildering. Said one executive with Gannett: "Now there are more questions and fewer answers. The online landscape has not cleared up a bit but rather has gotten a lot more complicated."[52] In deciding on a potential partner, newspaper companies wanted answers to a variety of questions: Would it be better to align with a company offering a familiar advertising model or with one demonstrating the promise of cutting-edge technology? Should the decision be based strictly on expenses or was the timing right to worry less about cash outlays and more about the potential to share in significant revenue? Would a better partner come from the computer industry or would it be one owned by a prominent global media company?

It is not hyperbole to suggest that newspaper companies agonized over these choices. Executives formed partnership selection committees, attended conferences, met with multiple potential partners, assigned dedicated personnel to study the technical differences, and hired new executives and consultants to sort out the best possible partner choice. For its part, the industry's trade association—the NAA—promoted an educational agenda aimed at assisting newspaper managers who were wrestling with these partnership decisions.

The following descriptions of five services were derived from a *Presstime* planning guide published in late 1994. These descriptions—presented alphabetically—illustrate the range of proprietary online services from which newspapers could choose a partner. These description excerpts are included because they provide insight into how these services were viewed by the newspaper industry's trade association, including some of the attributes that were deemed to be most important:

AOL:

Its mind-boggling expansion during the past 12 months—quintupling its membership with more than a million new subscribers—has made AOL the darling of both Wall Street and Infobahn newbies. The simplicity of its point-and-click interface . . . lured many new users and convinced several newspaper companies to cast their lots with AOL.[53]

CompuServe:

Newspapers desiring an inexpensive experiment with online publishing through a commercial service that still reaches millions of people worldwide should consider CompuServe . . . newspapers on CompuServe are probably the only ones partnering with a commercial service that already make a profit.[54]

Delphi:

Fans of "Beverly Hills 90210"—exactly the young consumers newspapers long to attract—have been following their TV idols lately to an information-highway address that is seeking newspaper partners: Delphi Internet Services Corp. On several November episodes of the popular Fox TV network show, the college-age characters took to their PCs to dial into Delphi, the online service that Fox owner Rupert Murdoch bought in 1993, and the Delphi logo was repeatedly conspicuous on their computer screens. Such tie-ins with other media suggest the advantage of linking with a global information company like Murdoch's News Corp.[55]

Interchange Network:

Interchange Network Co., designed by former computer-magazine magnate Ziff-Davis Publishing Co., is gaining partners by being the first commercial service to offer content providers "third generation" publishing tools . . . a software platform that offers powerful searching capabilities, hypertext links and other state-of-the-art features. . . . The risk in partnering with Interchange, however, is the uncertainty that surrounds the . . . company's future.[56]

Prodigy:

Prodigy is the only service that sends ads to customers along with any information they request. The ads take up the bottom portion of the user's computer screen, much as newspaper ads traditionally run at the bottom of pages. Also, until the past few months, when most commercial services began upgrading their software platforms, newspapers considered Prodigy's graphics and color-heavy interface the best bet for attracting advertisers to online ads.[57]

These brief excerpts highlight the issues that ranked highest in terms of publisher concerns: audience size, profitability, promotional opportunities, financial stability, and commitment to developing the advertising market. Newspaper companies, in their final analysis, gravitated toward AOL, CompuServe, and Prodigy. CompuServe was the choice of many middle-market newspaper companies where the requirement for the lowest-cost option ruled the process. A Gannett executive explained this rationale: "Some markets simply won't support online newspapers in any significant way. Under those circumstances, it makes sense to go with CompuServe and a very modest model."[58] A business development executive with the *Los Angeles Times* concurred: "If you're a mid-sized market, then you can afford to be casual. But not if you're in a large market with competition around you, with major players—cable companies, telephone companies—eyeing our revenue streams hungrily."[59] In these larger markets, however, there was not a consensus regarding a single best partner. The publishers who went with one of the major players were seen as aligning with market momentum in the case of AOL or advertiser acceptance in the case of Prodigy.

Presstime published an extensive review of AOL and Prodigy, attempting to explain both services from the perspective of a consumer user for the trade publication's audience of newspaper executives. The publication concluded that AOL's "virtue is its greater ability to search databases." Prodigy, however, had several "virtues we newspaper people like" including "ease of use and bright color graphics and advertising." The publication declared Prodigy's "interface easy to navigate, the graphics fun and the advertisements useful and unobtrusive."[60] From a negative perspective, the reviewer found that AOL's search features were limited, while Prodigy's graphics-rich interface could be slow to render on a computer screen and that its software sometimes interfered with other computer programs.[61] In the end, most partnering decisions were based on achieving a level of comfort with the online service and determining that the newspaper and the online partner could co-exist. To provide further insight into how these partnerships were developed, the chapter explores in more depth the newspaper industry's relationships with the two companies that emerged as the most prominent partners: AOL and Prodigy.

AOL

Tribune Co., the parent company of the *Chicago Tribune*, led the newspaper industry's relationship with AOL. Tribune had acquired a minority ownership stake of about 11 percent in AOL in its formative stage and aggressively pushed its services.[62] Although AOL was creating a nationally branded service, Tribune embarked on a plan to use its newspapers to create local affiliates within the

AOL service. Tribune later helped fund and establish Digital City as a separate, jointly owned business intended to exploit local market opportunities.[63] Tribune deployed classified advertising and online shopping services as part of its local affiliation strategy and worked to develop a transactional model in addition to news and information. *American Journalism Review* described Tribune's approach:

> Tribune decided to focus on its local markets rather than chase the dream of national supremacy on some lane of the information highway. The company also decided it couldn't compete with phone companies and other players in the big bucks efforts to build distribution systems on that highway. Instead, it saw its future in creating content and using technology to find new and profitable ways to sell that content to consumers.[64]

Tribune executives said they were attracted to AOL as a partner because of the vision it brought to the market. "We feel AOL is on the cutting edge as a marketing company in terms of how these services will develop," said the Tribune executive in charge of managing the partnership.[65] A key component of AOL's marketing strategy was to convince consumers that the service was simple to use. Although this marketing message was directed at the service's end-users, it also resonated with many of the newspaper executives responsible for selecting an online partner.

For example, a key decision maker at the New York Times Co. said AOL offered newspapers the best access to the online services market because of its simple design. "It was the best and easiest because you didn't need a manual to figure it out," he said.[66] For the newspaper companies engaged with AOL, the concept of simplicity also extended beyond its design to encompass broader issues. These companies were complimentary of AOL's willingness to cooperate regarding the complex issues involved in launching newspaper services within the platform even as the company struggled with its own growth pains.[67]

An executive with Knight-Ridder's *San Jose Mercury News*, which reached an agreement to deliver a full-text version of this daily newspaper via AOL, said he was "very glad to have started with AOL," adding that it had "taught us a lot about the online world."[68] Given its history with Viewtron, Knight-Ridder executives felt they had to choose an online service partner carefully and present the company's return to the consumer online services market to investors in a way that reflected a methodical, long-term approach. According to the editor of the *Mercury News*, AOL represented the best platform to accomplish this goal.[69] Careful to avoid industry criticism about unfounded expectations and a repeat of the Viewtron experience, this Knight-Ridder editor intentionally set the bar low for the partnership with AOL:

> We're in a continuum of long, slow change and innovation. People aren't going to flock to the electronic newspaper instantly. We need to use the printed newspaper to lead people into the new form, and this [agreement with AOL] is Step 1. . . . We don't know what the economics are, but we created a low-enough cost structure that it won't take very much to break even.[70]

For the newspaper companies engaged with AOL, the platform represented a simple, straightforward approach to the electronic services market. The publishers also recognized value in a marketing effort that propelled the service to three million U.S. subscribers by 1995, which was a growth trajectory that allowed AOL to tout a year later that it had become the "first billion dollar interactive services company."[71] However, many of the newspaper companies that did not affiliate with AOL indicated that they were seeking a partner with an approach that relied on advertising for the primary source of revenue.

AOL's business model relied primarily on subscription revenue and was, as a *Newsday* executive described, "the reverse of the newspaper model."[72] Indeed, at this point in the business's evolution, AOL executives viewed advertising very differently than newspaper publishers did. Advertising was treated as transactional content and relegated to a portion of the platform so that it would be "unobtrusive." One of AOL's key executives explained the company's rationale:

> We don't think advertising works in this new medium. Members are looking for information and a sense of community, and they want that in a safe and unobtrusive place. Typical advertising is designed to be intrusive.[73]

Newspaper publishers were intrigued by the prospect of improving on the ratio of revenue derived from subscriptions as promised by the AOL model. But they were principally in the business of selling advertising, which translated into a lukewarm response to AOL's advertising strategy.

By 1994 and into 1995, large newspaper companies were increasingly aware that advertiser interest in online media platforms was undergoing a transformation. A senior executive with a large New York advertising agency described the shifting attitude by observing that in fall 1993 advertiser "interest in online was zero," but less than two years later, "on a scale of one to 100, we probably see interest at 120 now."[74] As newspaper publishers sensed this shifting mood, their choice for an online service partner was influenced by the desire to capitalize on an emerging opportunity.

The next section explores how Prodigy attempted to exploit the market for online advertising during this period and, in the process, emerged as the principal online services partner for the newspaper industry.

Prodigy

Due in large part to its deep-pocketed corporate backers—IBM and Sears— Prodigy established credibility quickly in a marketplace full of lesser-funded competitors. Although Prodigy claimed to have officially launched in 1989, the business actually had its roots in a failed videotext project from 1984 known as Trintex.[75] By 1990, however, its owners had invested more than $600 million in retooling the business, and Prodigy was on its way to achieving status as an icon of its era.[76]

The Poynter Institute, a newspaper research and training organization, under-scored the importance of Prodigy within the timeline of the newspaper industry's involvement in electronic media by stating: "From an historical standpoint, Prodigy serves as a bridge from videotex to the new media projects of the 1990s."[77] Within the context of this book, Prodigy is important because it emerged as a strategic partner to several newspaper companies who were attracted largely because of its stated commitment to develop a revenue model based on online advertising. This is significant, of course, because advertising—not subscriptions—was the primary source of newspaper industry revenue. Advertising was the basis for a business model newspaper publishers understood, and one they wanted to translate into an online model that would work for them. Prodigy was seen by many within the newspaper industry as providing the best assistance in achieving that goal.

Soon after its launch, Prodigy had established relationships with more than 200 national advertisers.[78] Prodigy had made a concerted effort to appeal to these advertisers by embracing a radical departure from the technology norms of its day. Prodigy eschewed existing videotext technology in favor of a different presenta-tion architecture designed to render colorful, graphical elements on the computer screen much more readily than previous systems. Prodigy sought to appeal to advertisers and consumers by going beyond the typical user experience that was common up to this point. *Information Today* described Prodigy's motivation:

> Being attractive to users is what consumer videotex is all about. Videotex for business users can be simply utilitarian, but for the general public, it must be more than that. When appropriate, it should provide quick answers or meet users' immediate needs. But also, it should have some element of surprise, even fascination. Design elements have become all the more important now.[79]

Prodigy's efforts to create colorful, user-friendly designs were heavily influenced by its relationship with advertising agencies, especially J. Walter Thompson. Prodigy realized that most advertisers would be uncomfortable developing material for online services and would seek assistance from established agencies they trusted with television, radio, and print advertising.

J. Walter Thompson, a leading advertising agency at that time, had demonstrated a long commitment to working with potential new advertising delivery platforms. The agency even worked on advertising campaigns that had appeared on Viewtron and Gateway, but the failures of those services did not deter Thompson from an aggressive approach involving Prodigy. Working primarily with its auto industry clients, the agency opened an office in Detroit dedicated to creating campaigns for the online market, and it became "the first full-service agency for Prodigy." The agency was responding, its managers said, to what *Information Today* described as "two important transitions" occurring with online services: "The visual impact of online material is greatly expanding, and the . . . influence of advertisers in transforming videotex into something akin to another media."[80] Prodigy's relationship with Thompson and other agencies was not exclusive, but its leadership role in reaching out to the advertising community and actively catering to its needs is a major reason why it attracted the attention of the newspaper industry.

As AOL found early newspaper industry support from Tribune Co., Prodigy received an early and important endorsement from Cox Newspapers. Cox entered into a multi-newspaper deal and emerged as the industry evangelist for Prodigy, spearheading efforts to create a consortium for other newspaper companies to join the platform. Cox had a partnership underway with BellSouth's online gateway service as was discussed in the previous chapter, but Cox's president said he was attracted more to Prodigy because its approach so closely resembled a newspaper model. He specifically noted that Prodigy's design, which featured "substantial display advertising," was a close proxy for the format of a printed newspaper page.[81]

Cox deployed the Prodigy platform to put several of the company's newspapers online, including the *Atlanta Journal-Constitution*, the *Palm Beach Post* in West Palm Beach, Florida, and the *Austin American-Statesman* in Texas. During this process, Cox touted the Prodigy platform as a low-cost way for publishers to enter the re-emerging online marketplace. The newspaper company also boasted that Prodigy offered a technologically superior network that would serve newspapers well in local markets because its 130 local dial-in numbers across the country reduced long-distance access requirements.[82]

Prodigy initially launched with its own content staff, becoming the only one of the proprietary online services to promote original content as a service differentiator. Prodigy's staffing initiatives rekindled the newspaper industry's concerns that had surfaced in the 1980s debate with AT&T when newspaper executives fought to keep the telecommunications industry out of the news-creation business. As recounted in Chapter 3, AT&T tried to appease newspaper executives with assurances that the telecommunications giant had no intentions of creating news content, but only wanted the ability to freely distribute such content. Prodigy made no such concessions at the outset, thereby stoking newspaper industry concerns regarding a new breed of competition. Newspaper companies recognized the same kind of electronic threat that AT&T had represented, but Prodigy did not have the regulatory restraints imposed on the telecommunications industry.

Nevertheless, perceptions of competition gave way when both sides decided that cooperation represented a greater opportunity. As Prodigy embraced the notion of partnering with media companies, it abandoned its original content plans. After entering into deals with several media companies, including at least eight major newspapers by the end of 1994, Prodigy eliminated its internal content staff of at least 100 employees and turned solely to its new media partners for content. As part of this switch, Prodigy took steps to address some of the production issues and control concerns that newspaper companies had expressed.[83] A Prodigy spokesman described the company's transition as "moving toward an open-network strategy, allowing publishers to come on and maintain their brand identity, providing authoring tools to create the same look and feel as their newspapers."[84] Prodigy also teamed up with its partner newspapers and independent software companies to create new services and functionalities, including, for example, a tool that allowed users to select specific newspaper stories and assemble them into "personal on-line newspapers."[85] In addition to its advertising relationships, system functionality was also cited as a significant reason why newspaper companies chose Prodigy as its online partner.

The *Los Angeles Times*, for example, followed Cox onto the Prodigy platform,[86] creating TimesLink, which was described by an executive as "a powerful, local, online gateway to commerce." This executive maintained it was the "functionalities" of online services including interactive chat and message boards, more so than content, which attracted users and created community. "By drilling deep, we can begin to move the needle beyond the high-tech audience" to a mainstream audience, he said.[87] An executive with the *Tampa Tribune* also expressed the desire for pursuing a mainstream audience as a factor in selecting Prodigy. He explained: "If we had wanted to go with strictly the techie market, we would have chosen CompuServe or

America Online. But the feeling was that with Prodigy we could go get new people who had just purchased a computer for the first time."[88] It is difficult to grasp such basic concerns when viewing them through a contemporary lens, but consider that in 1993 less than 13 percent of U.S. households had a personal computer equipped with a modem capable of accessing an online service. Newspaper executives believed the market would expand rapidly, so aligning with an online partner was essentially placing a bet on the one considered to be the best horse in the race.

Transition to the Internet

Newspaper companies never had the chance to see how their bets with the proprietary online systems would pay out, because the unexpected emergence of the Internet disrupted the market. Even as newspaper companies were signing their deals with proprietary online service companies, the Internet was creeping into the market. The newspaper industry's ability to adapt would be challenged again as consumers began to embrace the Internet in the mid-1990s.

Newspaper publishers can be excused for not appreciating in 1993 and 1994 how disruptive the decentralized Internet would become. Although a few pioneering newspapers, including the *News & Observer* in Raleigh, North Carolina,[89] experimented with the Internet as early as 1994, most newspaper companies did not see the potential. They were not alone. Even the technology giant Microsoft is remembered as slow to grasp how transformative the Internet would become. As late as August 1995, Microsoft was just getting around to launching its own proprietary service, the Microsoft Network. Some newspaper executives fretted over Microsoft's entry as a possible new competitor, but a newspaper in Microsoft's backyard—the *Seattle Times*—announced that it would experiment with providing content on the new platform.[90]

With the Internet looming and a potential powerful new proprietary service backed by Microsoft entering the market, newspaper companies were again confronting a confusing and unsettled time. They once again faced critical questions about strategic direction. One executive describing this period said that "if the new media landscape is Oz, we are about a half a step down the yellow brick road."[91] The comment was prescient as the media landscape changed rapidly in the mid- to late 1990s, especially as the proprietary services morphed into Internet service providers before fading in importance.

In revisiting this period of transition, Reid noted that services such as Prodigy, AOL, and CompuServe "were flourishing" as the Internet appeared. Given the demonstrated popularity of these services, Reid questioned how the Internet was able to attract so many users so quickly. He wrote that the success of proprietary online systems "indicated an increasing interest in connecting and communicating among computer owners," but he observed that "while the commercial services helped satisfy that urge, none ever attained the kind of content and user growth momentum that the Web generated."[92] Reid's conclusion as to the reason for the Internet's ultimate success and the subsequent decline of the proprietary services is essential to understanding the riddle facing newspaper executives as they contemplated their next moves. Reid wrote:

> . . . each commercial service's subscribers and content were sequestered from the others'. Subscriber growth at CompuServe therefore did nothing to help Prodigy's content reach more of its natural audience, and growth in America Online's content did nothing to enrich the CompuServe or Prodigy experiences. Segregated and barricaded, the aggregate online population could never achieve the full benefits of mutual affiliation that open networks offer.[93]

Newspaper executives who had worked hard to hammer out their partnerships with the proprietary online services were glad to have found a role in the electronic marketplace that made business sense. Providing content within the gated communities of the proprietary online systems was comfortable. The Internet was not. Many in the newspaper industry initially believed that the open network of the Internet would lead to information anarchy. Without central control, how would consumers know where to turn? Without standards, how would advertisers ever trust the medium?

The newspaper industry's response to such questions is the focus of the next chapter. But it is important to note that there was no clean transition for newspapers from the period of proprietary services to the Internet era—no date on a calendar that marked a seminal event. Until 1995, the American newspaper industry had largely responded to the renewed possibilities of online media by forming partnerships with proprietary online service companies. These companies also did not anticipate the revolutionary impact that the Internet—and its content from the World Wide Web—would have on their plans and expectations for online media. Even as late as May 1996 when other services were migrating most services to the Internet, AOL said it would remain a proprietary service. "We have no plans to become a HTML shop anytime soon," said its head of product marketing.[94] Its newspaper partners continued to be an important source of content during the transition years, especially in areas that emphasized local, geographically focused content.

Nevertheless, as newspaper executives realized the staggering rate at which the Internet was achieving household penetration, they began to take their content there as well. Boczkowski summarized the period covered so far by this book—the 1980s to the mid 1990s—as a time that "American dailies tinkered with an array of alternatives to print."[95] He elaborated:

> . . . the 1980s was a decade of exploration of multiple technical, editorial and commercial options. While newspapers continued to explore these options during the first half of the 1990s, they progressively narrowed down their efforts around products delivered to personal computers connected to online services until they finally settled on the web circa 1995.[96]

Boczkowski's notion of an industry tinkering for a decade and a half perhaps implies that the efforts were half-hearted. Maybe they were in some cases, but that does not diminish the influence these projects had on the collective industry thinking about electronic distribution of news, information, and advertising.

Discussion

The brief period of concerted engagement with the proprietary online services can be viewed as the time when newspaper publishers turned away from their protectionist rhetoric of the late 1980s and became seriously engaged once again in efforts to exploit online media. Working with companies such as Prodigy and AOL gave newspaper companies a renewed sense of purpose in the electronic era and fostered optimism throughout the industry. While the earlier failures of Viewtron and Gateway had tarnished the industry's reputation, the work with the proprietary online systems restored it.

The fact that newspapers were active again in pursuing new technology represented a seminal change for an industry that had spent the better part of the 1980s fighting the telecommunications industry and allowing its reputation to be cast as a group of naysayers and obstructionists. The notion that the industry's future success would require conquering and exploiting new technology was once again a commonly held belief.

By 1993, it was no longer considered optional for a serious daily newspaper to be involved in the online world. This attitude shift is corroborated in an influential textbook on newspaper management that listed a key strategy for success as "joining high tech companies in jointly producing interactive electronic services or in other ways using new technology."[97] The key part of the strategy involved partnerships. Despite the renewed optimism, the skittishness about the cost of electronic information services lingered within the industry. It fostered a climate where few newspaper companies wanted to invest in online services independently.

Even so, the recognized shortcomings of the new proprietary online services— unproven technology and business models—would have been more than enough to dissuade newspapers from teaming up with them only a few years earlier. But the underlying climate had changed. *American Journalism Review* called it "not a revolution, but an evolution," adding that "recent shifts with the cable and telephone industries, the surge in popularity of personal computers, and a drawn-out economic recession that has sliced into advertising revenues have combined to force newspapers to look forward."[98] In looking forward, however, the newspaper industry envisioned an online marketplace that would largely resemble its offline world. In this market, large media companies centrally created the content, controlled its distribution, and relied on advertisers to pay for the majority of it. In this regard, newspaper companies understood the proprietary online services largely because they also relied on central control.

Chapter 5 continues with an examination of the newspaper industry's ongoing desire to participate in the online services market. The chapter tracks the newspaper industry as it transitioned from the proprietary services to the Internet. It explores the latter half of the 1990s as newspaper companies launched numerous Internet ventures and tried to adapt to an emerging media model that called for radical changes in the way newspaper companies preferred to operate. This book has shown that the newspaper industry often acted defensively and—after the early 1980s—with risk management concerns at the forefront of its strategies, but there was a résumé of online experience in place when the Internet arrived. The next chapter examines how newspaper industry executives thought they could use that experience and their local market dominance to exploit the Internet to their advantage.

5

The Emerging Internet Threatens the Established Publishing Model

The emergence of the World Wide Web during the time when newspaper companies were aligning their future with proprietary online services was unexpected and unsettling. *Wired* magazine's October 1994 article entitled "The (Second Phase of the) Revolution Has Begun" would have provided reason for concern for any newspaper executive who may have read its opening: "Don't look now, but Prodigy, AOL, and CompuServe are all suddenly obsolete—and Mosaic is well on its way to becoming the world's standard interface."[1] Newspaper executives, just getting comfortable with their new online strategy and the ensuing partnerships, were forced to confront this new development. They clearly were confounded by the accelerating pace of change and unsure of how best to address their industry's role in online media in the mid-1990s.

This chapter explores the newspaper industry's reaction to the World Wide Web as its development established the Internet as a full-fledged content distribution platform. Initially, newspaper executives were skeptical. After all, the giant Microsoft was seen by the newspaper industry as a technology bellwether, and its actions during this period did not portend the Internet to be as revolutionary as some commentators were suggesting. Microsoft, for instance, largely followed the proprietary model when it launched the Microsoft Network (MSN) in August 1995,[2] which was ten months after the *Wired* article pronounced such endeavors obsolete.

Once the movement of major content providers to the Internet began, however, the newspaper industry represented significant activity. This chapter examines several projects undertaken by the industry that illustrate its response during the period that led up to the Internet industry's financial bubble collapse in 2000. In explaining how this profound transition unfolded, this chapter looks first at the development of Mosaic, which was critical to the overall acceptance of the Internet as a mainstream platform. The focus then turns to the newspaper industry's migration to the web, and highlights the NCN initiative. The demise of NCN is examined, including retrospective comments from participants and observers regarding what its failure represented in terms of the newspaper industry's ability to exploit the changing marketplace. The industry wrestled with numerous issues involving content, structure, and business models. There is particular emphasis given to the classified advertising component of the newspaper industry's business model and how emerging businesses on the Internet began to alter the dynamics of that market.

The Mainstream Internet

When newspapers first encountered the Internet, wired households in the U.S. were far from ubiquitous. Only 11 percent of the country's households owned modems in 1994. The number grew to 26 percent by 1997 and to about 33 percent by the end of the decade.[3] The overwhelming majority of households during this period did not own the equipment necessary to access the Internet. Therefore, newspaper executives were not reacting to the Internet's market penetration, but rather to its growth trajectory. In looking back over the Internet's development, many assume that its dominance was recognized as inevitable from the outset. For those with only the perspective of the mid-1990s, however, the future of the Internet was only speculation.

AOL, Prodigy, and CompuServe were attracting enough mainstream audiences to entice Microsoft into the arena for proprietary online services, but none of these businesses contemplated an Internet-based service at this juncture. The conversations about the Internet during this period were found in the academic and technical communities and among some early adopters. As the conversations about the Internet gained momentum, they were joined by entrepreneurs and venture investors, which formed a collective that would become known as the *digerati*.[4] The rhetoric emanating from these early conversations foreshadowed why the Internet would be different from the online information services that preceded it.

Mitchell Kapor's essay published by *Wired* in the summer of 1993 captured these sentiments in explaining how this new group of people would deliver on the promises of an Information Society where others had failed:

> This dream has been promoted extensively, but until recently little visible progress has been made toward its realization. In the past, political gridlock has snarled telephone companies, newspaper publishers, cable television operators, and other potential players in lengthy and fruitless congressional and court battles. A justifiable cynicism developed to fill the gap between vision and reality. Meanwhile, the pioneers of the computer-mediated communication networks collectively referred to as cyberspace are not willing to wait. Employing whatever tools they can find, they are constantly pushing the techno-cultural envelope. Life in cyberspace is often conducted in primitive, frontier conditions, but it is a life which, at best, is more egalitarian than elitist, and more decentralized than hierarchical. It serves individuals and communities, not mass audiences, and it is extraordinarily multi-faceted in the purposes to which it is put.[5]

The essay reflects the notion that traditional media, including newspapers, had not done enough to develop electronic information services. The implication was that the established players had failed to exploit the technology available to them, and therefore, it was time to allow a new breed of businesses—those not wedded to models of control—to establish a new operating model.

Traditional media, however, saw no reason to simply step aside. The newspaper industry specifically forged ahead with plans based on its long-established media model that relied on creating content centrally and distributing it en masse from the point of creation to multiple end-users. Newspaper industry executives believed their view of the electronic information market had been corroborated by the success of commercial databases, which operated on a similar model. In these services, data was collected in large central repositories and delivered through proprietary computer connections to customers who subscribed. In both the media model and this database model, centralized control of content and distribution was the fundamental element.

The Internet, however, was not conceived as a publishing tool in the traditional sense. Rather, its roots were in the research labs of the nation's government, defense, and academic institutions. An original purpose was for researchers to collaborate more readily by sharing information; therefore, its design allowed content to flow from any point on the network. There was no central authority to control what was shared, how much was shared, or when it was shared. The Internet was inherently interactive because it was specifically designed to foster the sharing of data and to facilitate communication about the data that was being shared. Traditional publishing, on the other hand, was designed to empower the owners of content who intended to profit by controlling its dissemination.

This fundamental difference empowered the users of the Internet in ways that traditional publishing was not designed to do. It gave users control over the content they consumed, while traditional publishing restricted control. For years, however, the Internet's users were limited to technically savvy individuals who understood how to write and communicate in complex computer languages. The Internet did not become a useful consumer platform until an application was developed that hid the computer complexity behind a simple graphical interface. In the following material, this chapter explores the significance of this development and examines the implications it had for increasing the popularity of the Internet and the ramifications for the newspaper industry and its traditional publishing model.

The Game-Changing Mosaic

Online magazine explained that the mid-1990s represented a transformative point in media history because of the functionality unleashed by the World Wide Web:

> . . . it was the genius of linking pages and hyper-jumps in digital space that turned the trick. The next revolution was unleashed. The Internet became the computing infrastructure, the definitive metaphor for electronic information. . . . In 1987, online included the Internet. By 1997, online was the Internet.[6]

There is little attempt in current discussion to distinguish the Internet from the World Wide Web, but at the outset there was a clearer understanding of the Internet as the network and the World Wide Web as its primary content layer. The distinction began to evaporate with the introduction of Mosaic, a software application that provided a graphical interface for finding and navigating through content using the Internet.

Hypertext had been around for years, but Mosaic brought its potential to life in a way that general computer users could understand and use. In doing so, the application triggered a rush to create content and fostered the notion of the Internet as a media platform. An early user of Mosaic recalled her experience as transformative: "For many of us involved in early online and interactive publishing experiments, Mosaic, or as I liked to refer to it in a time of less potent bandwidth, 'The Little Browser That Could,' changed our world."[7] In the case of Mosaic, descriptions such as revolutionary, world-changing, and transformative were not hyperbole. CNET News, for example, explained the significance of Mosaic:

> . . . the modern concept of the Internet would not exist if the browser had remained in the exclusive realm of academia. . . . Mosaic transformed the Internet from the esoteric province of researchers and technophiles to a household appliance, creating a multibillion-dollar industry and changing the way society works, communicates and even falls in love—in short, affecting nearly every facet of life.[8]

Before the creation of Mosaic, the Internet and World Wide Web were not topics of mainstream discussion. The *New York Times* published its "first article about the Web" on December 8, 1993,[9] and described Mosaic as "a map to the buried treasures of the Information Age." The article explained that Mosaic did not provide connections to the Internet, only a more convenient way of finding information once a person was connected.[10] Effusive in its praise, the article noted that Mosaic's "many passionate proponents hail it as the first 'killer app' of network computing—an applications program so different and so obviously useful that it can create a new industry from scratch."[11] Many would argue that Mosaic did just that. From the outset, it was treated as an exciting new software application, but in less than two years it was a global sensation as it enabled the Internet—as we understand it today—to exist.

Its significance also stems from the role it played in igniting the Internet as a business phenomenon. Mosaic's success gave rise to the Internet as a breeding ground for innovation and a lucrative place for venture capital investors to find opportunities. Tim Berners-Lee, a British researcher, is credited with establishing the technical protocols that created the World Wide Web as a way of organizing information. Mosaic, however, represented a "better way of displaying the information that the Web had so brilliantly organized."[12] Mosaic represented such a powerful innovation that its creator, Marc Andreessen, a student at the University of Illinois, used it as the foundation for launching Netscape.[13] The company was also a game-changer, representing the first in a series of Internet-based start-up companies that altered the face of American business during this period.

A New Business Climate

In August 1995, less than two years since its creation, Netscape went public through a landmark stock offering.[14] Despite losses that had reached nearly $13 million, Netscape soared to a market value of $2.7 billion on its first day of trading.[15] The unprecedented event on Wall Street touched off the investment frenzy surrounding Internet companies in the latter half of the 1990s. The losses notwithstanding, the success of Netscape's stock offering was supported by the perceived potential of its flagship product, the graphical browser that had its roots in Mosaic. The exuberance investors showed for money-losing Internet start-up companies transformed the capital markets during this period. The events confounded executives in many industries, including the newspaper business. They were dismayed by the magnitude of capital investment that was flowing freely to companies that were unproven, had posted only losses, and were operating in a market with no established history.

One newspaper industry consultant observed that the financial markets were rewarding companies that brought innovative products to markets without regard to their business pedigree. The consultant stated: "Content kings are now vulnerable," adding that "newspapers bring capability and capital . . . but they won't be chosen over the kid down the street who knows how to make and use those tools. We need to get to know those kids."[16] The implication of the statement was that newspaper companies were unlikely to be innovative enough on their own to compete in the new Internet market. Previously, traditional media companies such as newspapers were thought to be important players in electronic information because of the inherent value of their existing content. The early developments of the Internet, however, demonstrated that functionality, interactivity, and community building could perhaps trump the importance of content held by the traditional media, a topic explored later in this chapter. However, the newspaper industry was concerned about other shifts in the media landscape beyond the emergence of the Internet during this period.

Newspaper executives were forced to consider their strategic decisions in the context of an escalating pace of mergers occurring among their more traditional electronic rivals—radio, broadcast television, and cable television—as well as among the phone companies. Chan-Olmsted found that among those industries, the number of mergers and acquisitions—transactions valued at more than $5 million—soared from 68 deals in 1991 to 204 deals in 1996.[17] The period was marked by a surge of activity in 1996 spurred on by new telecommunications legislation that sought to bring policy in line with the new technologies. The legislation brought the specter of more wholesale changes to the media landscape. As Chan-Olmsted concluded:

> The sweeping changes in telecommunications regulation signed into law in 1996 by President Clinton set the stage for a new era of strategic alliances among communications services companies in the country. By striking down the rules

that have prohibited cross-ownership between telephone and cable companies, limited broadcast station ownership by a single entity, and separated local and long-distance telephone service providers, the new regulatory environment will likely foster the convergence of broadcasters, phone companies, and cable TV services. . . . Convergence through M & A seems to provide the best opportunity for companies to accelerate the implementation of new technologies and at the same time capture a developed customer base.[18]

The newspaper industry was at a crossroads. The emerging Internet had many executives questioning whether their alliances with the proprietary online services were sufficient. This new concern contributed once again to a fear of being left behind. When combined with the concomitant changes in telecommunications regulation, the newspaper industry was compelled to do more in the electronic arena than it had since the early 1980s.

Boczkowski wrote that "caught in the middle of these rhetorical and policy developments, many newspaper people imagined the Internet and related techno-logical changes to be tied to dramatic transformations in their own industry."[19] As newspaper executives began to accept the Internet as a paradigm shift of immense proportion, the industry mobilized in ways that were unprecedented. The news-paper industry committed to new investment in technology, new hiring, and a new industry-wide effort to collaborate as the industry collectively embraced the Internet as its path to the future. The remainder of this chapter critically examines specific aspects of the newspaper industry's Internet effort during the latter half of the 1990s.

Newspapers Mobilize for Changing Market

Newspaper companies committed to their Internet efforts amid a sense of urgency that had been missing from the industry's electronic endeavors for some time. As the head of the industry's trade association put it: "if we sit back and [it] ends up becoming a major information channel, we'll never catch up."[20] Many newspaper company leaders expressed confidence that the industry's earlier shortcomings in online media could be overcome. An industry leader stated: "It is informa-tion *processing* that is newspapers' greatest strength," adding that "newspapers must understand that what they do best—gathering, packaging and distributing news and information—is much more than ink on paper."[21] Another executive's comments also are illustrative of the strident views expressed about the ability of newspapers to become formidable competitors in the new arena:

> Without a doubt, newspapers are best positioned to thrive in the electronic future. We already have strong brand recognition, established credibility and strong relationships with our customers.[22]

In using such attributes to describe the newspaper industry's strengths, leaders were seeking to reaffirm the value in their existing franchises and explain how that should differentiate newspaper companies in a new media market where upstart competitors had no history to build upon. Throughout this early Internet period, industry leaders also wanted constituents—readers and advertisers—to view newspaper companies as being in control of the rapidly evolving events.

The industry's trade association played an important role in developing this message. For example, when the association launched its own website in the summer of 1995, a statement described newspaper companies as "in the vanguard, exploring how this new medium can help translate their information and community-building franchises to cyberspace."[23] The language used, such as "vanguard," illustrates how the industry promoted its role as a leader in the emerging Internet-based media industry. Such efforts were effective in helping the newspaper industry shed the negative perceptions associated with its lobbying stance against the telecommunications industry and its early failed videotext projects.

As newspaper companies began to launch Internet editions, the industry received recognition for pioneering work. *MediaWeek*, for example, noted the newspaper industry's tenacity for pursuing electronic media projects:

> Newspaper executives don't give up easily. In the 1980s, many large dailies considered videotext a can't-miss technology for an industry that was suffering from stagnating circulation and rising newsprint and distribution costs. A decade later, a confluence of those same problems, and the emergence of modem-equipped personal computers, has newspaper executives talking electronic delivery again, substituting the Internet for failed videotext experiments like Knight-Ridder's Viewtron and Times Mirror's Gateway.[24]

The industry's public relations efforts would have had little effect in changing perceptions had newspaper companies not responded to the Internet in a significant way. But industry statistics show that the rhetoric was supported by significant activity.

About 60 newspapers in North America were publishing an online version of some kind by 1995 via a website, a bulletin board service, or through a proprietary service.[25] By the end of the following year that number had climbed to 230 papers online.[26] In May 1998, less than two years later, the number of online newspapers had exploded to 1,749, and most all of them were using the web rather than a proprietary service.[27] Newspapers placed so much emphasis on their web presence by this time that most other electronic publishing activities, such as audiotex, fax services, and partnerships with proprietary online services, were abandoned or curtailed. Knight-Ridder also closed its information design laboratory in 1995, which had been working for three years to develop an electronic tablet newspaper. In announcing the closure, the company said it planned "to concentrate our resources on Internet and online publishing."[28]

Boczkowski observed the newspaper industry's years of experimentation led to its adoption of the Internet in 1995 as its electronic platform of choice. As he surmised, had newspapers not experimented with earlier forms of electronic media, the transition to the web would have unfolded very differently than it actually did. The efforts newspaper companies had put into electronic distribution made it easier for them to accept the Internet as the next step in a development process that was already underway.[29] One research team wrote that projects such as videotext and commercial online databases were all "early forms of online publishing" that "in some way shaped the path of Internet publishing."[30] These efforts, chronicled throughout the preceding chapters of this book, led many newspaper executives and researchers to believe that the industry's experiences had produced knowledge that would be transferred to the Internet endeavors.

Lee, for example, observed that "early trials and offerings of electronic newspapers have provided some useful insights valuable to the development of newspapers on the Internet."[31] Lee wrote the early online projects taught newspapers that they should be concerned with audience critical mass, onscreen readability, intuitive navigation, and interactivity.[32] Even though the newspaper industry entered the Internet era with this reservoir of online experience, its many critics contend most of the lessons learned went unheeded. In its subsequent sections, this chapter critically examines how the newspaper industry approached a number of operating decisions related to the emerging Internet marketplace. These decisions included a range of issues such as business partnerships, production processes, and competitive responses.

In the early going, newspaper companies were willing to experiment and approached the Internet as an opportunity. They entered business arrangements that were considered radical by the newspaper industry's previous standards—creating partnerships and making investments that challenged the industry's cultural make-up. The most notable example is an industry consortium called the New Century Network (NCN) that at one time represented the industry's best idea for achieving success in the Internet era. The creation and eventual failure of this effort was a milestone event during the newspaper industry's early response to the Internet.

New Century Network: An Aggressive Move

The newspaper industry's most aggressive attempt to gain an upper hand at the outset of the Internet era was in creating NCN, a consortium of several major newspaper companies. Its mission was to lead the industry's transition to the Internet, providing a framework for the nation's daily newspapers to establish a consistent online presence. The consortium's founders at its launch in April 1995 included Advance Publications Inc., Cox Newspapers Inc., the Gannett Co., Hearst Corp., Knight-Ridder Inc., the Times Mirror Co., the Tribune Co., and the Washington Post Co.[33] A few weeks later, the New York Times Co. joined, stating that once it had reviewed the membership proposal, the consortium was determined to be "a good fit for the company."[34] With the New York Times Co. on board, the ownership of the consortium represented more than 200 daily newspapers, and it planned to have 75 of those papers operating within the framework of NCN within two years.[35] The timeframe reflected a cautious and methodical approach, but one that underestimated the speed at which the Internet would develop.

A cornerstone of NCN's mission was to bring order to the Internet, and by doing so, allow the newspaper industry to exercise significant control over the emerging platform. NCN was founded during a narrow window in the Internet timeline that came after the creation of Mosaic, but before there was a clear sense of the Internet's development trajectory. How the Internet would develop was an unknown, but newspaper companies believed their local market presence and relative economic strength would give them enormous influence in determining how the technology would be deployed. In these early years, newspaper executives saw the emerging Internet as simply another electronic connection to the home. Although the Internet's presence in the media marketplace was unforeseen, newspaper companies did not react out of fear in this case. Rather than a defensive move, NCN represented a bold offensive strategy for a newspaper industry that was uncomfortable with the early anarchy of the web. Instead of waiting for order to emerge, many executives within the newspaper industry believed NCN could translate the industry's clout into standards regarding how the emerging Internet would operate as a media and advertising platform.

Newspaper executives especially wanted NCN to develop platform standards for advertising. Many newspaper companies were reluctant to move forward with

Internet initiatives in the absence of known advertising standards, including things such as common sizes that would allow agencies to place the same advertising across multiple websites. In discussions about how to bring standards to market, NCN once considered developing its own browser software.[36] While creating a new browser was deemed impractical, NCN was guided in its early months by its standards focus. One executive active in the NCN creation stated: "standards will allow newspapers to take precedence."[37] The newspaper industry's belief that it could influence the direction of the Internet's development through NCN underscores the importance of this initiative within the industry's early Internet efforts.

The NCN initiative surprised some of the financial analysts who followed the newspaper industry, but the concept was generally well received. An analyst stated: "It's surprising, because the industry doesn't often get together like this, but it makes a lot more sense than somebody trying to do something on their own."[38] The word "portal" had yet to be applied to the concept of an Internet destination intent on aggregating users, but that describes the early vision for what NCN could become. The initial business model was based on creating a newspaper industry hub, linking "member publications through common search, financial-transaction and advertising engines."[39] It also intended to provide a standard technical platform for delivering ancillary services such as home shopping, electronic mail, chat services, and discussion boards.[40] Nevertheless, the *New York Times'* description of how the consortium would operate seems quaint given an understanding of how the Internet evolved:

> Each newspaper would be able to create its own "look" on the Internet and retain its own reporting and editing staff for local news. But each would also be able to link with other newspapers, allowing, for example, a reader of the electronic *Austin American-Statesman* to buy a single copy of the electronic *Des Moines Register*, if he or she were particularly interested in news developments in Iowa that day, or to buy a subscription to the electronic *Houston Chronicle*.[41]

NCN's primary content activity was to aggregate news content from newspapers around the country, but to accomplish this in such a way that local branding remained intact. NCN's model reflected a television model. NCN would operate as the national network, while each newspaper would serve as a local affiliate.

To pay for itself and to share revenue with its newspaper affiliates, NCN planned to establish an interactive advertising network that could leverage the capabilities of the medium. NCN's interim chief executive explained: "On-line media are response-driven, and that means that the advertising content itself has to be entertaining and informative. This is the only medium that allows a user to make an immediate impulse request or purchase, so every package must have that capability."[42] NCN projected that advertisers would begin transitioning from print to the Internet when at least 25 percent of consumers—a critical mass— began regularly using the Internet. In 1995, NCN believed that threshold would be met by the end of the decade. The interim chief executive told his colleagues: "This means we have a five-year window, which is not long, to become competent in the intricacies of developing new interactive content and figuring out which advertising models best suit those new applications."[43] While there was a sense of urgency implicit in those remarks, the timeline was entirely out of line with how the Internet was actually developing. Nevertheless, they illustrate that the industry was poised to act. But agreeing on the details of how to act proved elusive for the newspaper industry.

Even though NCN had been created to embrace change, many newspaper executives remained ensconced in their risk-averse, conservative approach. Therefore, creating a vehicle for change did not immediately translate into support for action. NCN was created as an offensive organization with ambitious plans to bring advertising standards and uniform media processes to the Internet, but translating that into an actual operating plan proved difficult to achieve. The *New York Times* wrote that ". . . a partnership of large newspaper companies formed [NCN] to bring the country's dailies into the age of the Internet. Then they sat down to figure out what that meant."[44] The ensuing internal debate about how the goals for the consortium should translate into actual products and services lasted too long for many newspaper companies. The *New York Times* observed that "the project was quickly overtaken by the speed of Web development among papers."[45] Also, the overall pace of Internet development led many newspaper executives to concede that their industry's ability to influence Internet standards was not as immense as originally thought. As this realization took hold, the original dream of NCN leading a united newspaper industry into the new Internet-based market began to fade.

In less than six months from its inception, the first major crack in the unified effort appeared. On October 17, 1995, six major U.S. dailies announced the founding of CareerPath.com, touting an ambition to "offer the most comprehensive listing of jobs on the global computer network."[46] The six papers—*Boston Globe, Chicago Tribune, Los Angeles Times, New York Times, San Jose Mercury News,* and *Washington Post*—were owned by the companies who had been part of creating NCN. CareerPath said it intended to "coordinate" its activities with NCN, and an NCN statement described CareerPath as "an example of the type of initiative we envisioned when NCN was formed," adding that "our affiliate strength will give us the opportunity to create, promote and provide easy access to a uniquely powerful interactive employment marketplace."[47] Nevertheless, CareerPath represented a significant breach of industry solidarity that NCN was thought to have represented. Knight-Ridder Inc. also stepped outside the NCN boundaries to create a venture with Landmark Communications to provide Internet services for newspapers, an effort in which it hoped to attract 100 daily newspapers as clients. The joint venture said it intended "to complement—not compete with—NCN," but it was not specific about how it intended to do that.[48]

By the end of 1995, NCN's mission was muddled. NCN's owners appeared uncomfortable casting their entire Internet lot with their industry brethren, but they did not shut down the consortium at that time. Instead, NCN brought in new leadership from outside the newspaper industry in the person of Lee DeBoer, a television industry executive most recently with HBO.[49] Under DeBoer, NCN attempted to regroup by launching a content syndication product called NewsWorks, which was described by its editor as follows:

> Imagine having a personal news assistant who reads every word published every day in more than 125 newspapers . . . and then gives you exactly the news and information you need and want. That's what NewsWorks does.[50]

NewsWorks was positioned as network service and emphasized its role as a content aggregator. The service touted the scope of its offering by explaining that its content was drawn from newspapers across the country and represented the collective effort of more than 25,000 journalists.[51]

CNET called NewsWorks one of the newspaper industry's "most aggressive efforts" to counteract a "sleepy response to online advertising."[52] Although the size and scope of the offering was impressive, reviews of its execution were mixed. *CNET* explained that NewsWorks did not post news articles as stories broke, opting instead to aggregate them after they had appeared in the affiliate's print edition. "That may make them too stale for some Netizens' appetites," CNET wrote.[53] Another analyst concurred, stating that NewsWorks represented "one more stage of recycling the same old news."[54] However, *Information Today*, which catered to research professionals, called NewsWorks "one of the most powerful news research instruments on the web." The review favorably compared NewsWorks to commercial services such as Dialog and NEXIS. Given that NewsWorks was advertiser-supported and free to users, the magazine said that choosing NewsWorks over an expensive commercial database was "a no-brainer," adding that "nowhere are the peculiar economics of cyberspace demonstrated more sharply than with NewsWorks."[55] In NewsWorks, NCN had created a robust research tool, but it reflected the newspaper industry experiences with commercial information business models more so than consumer experience.

Knight-Ridder, for example, was a significant operator of commercial database services, and it had purchased Dialog in 1988.[56] Other newspaper companies had distribution agreements with commercial database companies as well. It was believed that such experience would be useful to newspaper companies as they moved electronic services to the Internet, but NewsWorks demonstrated how difficult it would be to translate the learning achieved in a commercial setting to the consumer Internet.

Information Today wrote that NCN needed to deliver "highly attractive locally oriented services" and "convince local advertisers that local services are a legitimate media buy," but the magazine concluded that the newspaper industry's "dismal record of previous attempts" made success difficult to predict.[57] NCN's prospects were further diminished when key companies decided against the platform. Although the New York Times Co. and Gannett were NCN owners, neither of their flagship newspapers—the *New York Times* or *USA Today*—ever affiliated with NCN,[58] rejecting the common platform in favor of creating their own web initiatives.

By January 1998, NCN had all but abandoned NewsWorks. It eliminated 10 percent of its staff and said it was shifting to a model of delivering content via HTML-based email on behalf of its affiliates.[59] That effort never launched, however. In the following month, NCN withdrew from providing news content, stating in a press release that NCN would focus only on delivering advertising: "Our newspaper affiliates create a vast amount of rich interactive content every day. The network can add the most value by creating marketing solutions for national advertisers built around affiliate content," the statement read.[60] The repeated attempts to change strategy could not overcome a deeply divided ownership structure.

Less than two weeks after abandoning NewsWorks, the entire NCN business was closed. The board of directors representing NCN's newspaper ownership issued a statement on March 10, 1998, stating that it had "voted to dissolve the partnership." The statement added: "The challenge of concurrently finding a common strategy for NCN . . . has proved too daunting.[61] The creation and ultimate demise of NCN may be only a footnote in the overall history of the Internet, but it was a watershed event when viewing the development of the Internet from the perspective of the newspaper industry. The creation of NCN underscored the newspaper industry's recognition of the Internet as an important new development in the media landscape, but its failure reinforced the perception that the newspaper industry would struggle to compete in the new market.

Therefore, examining its brief history provides insight into the industry's collective thinking during this period. An industry executive observed that NCN marked a time when "a bunch of newspaper companies got together and imagined an online future in which newspapers would be key players," believing that they could build "the definitive news resource on the Internet."[62] In retrospect, NCN illustrates how internal arrogance and naivety regarding external forces affected the newspaper industry's early response to the emergence of the Internet.

For example, 18 months lapsed before NCN launched NewsWorks, its first major product. The long delay demonstrated a failure to understand and react to the speed at which the Internet was developing. An industry participant recalled it as a time when "technology was evolving faster than anyone's business vision." He added:

> I remember seeing one of the early NCN definitions in the form of a request for proposals to provide infrastructure for the network. The idea was for a federation of closed, paid-access websites, where you could navigate from site to site on a single membership pass. . . . By the time the RFP made it through the fax machine (yes, it was faxed) the idea was obsolete. Switch gears: Open portal. Switch gears: search engine. Switch gears: ad network. None of it worked and some players were left angry and bitter.[63]

By the time NewsWorks was in the marketplace, the industry support for the consortium had diminished. Newspapers were concentrating on their own independent websites and were openly questioning the need for a national network. The following anecdote illustrates how indifferent the industry became of its own creation.

In reporting on an NCN reception at a newspaper industry convention, *BusinessWeek* wrote:

> A thousand bottles of champagne emblazoned with "New Century Network: The Collective Intelligence of America's Newspapers" awaited the hordes expected to come toast the watershed new-media joint venture. . . . When fewer than 100 people showed up, Chief Executive Lee [DeBoer] made an abbreviated speech before retreating.[64]

BusinessWeek recalled the incident as NCN's "first public humiliation," but added that it was "only one in a series of blunders that culminated in the company's abrupt shutdown."[65] The magazine quoted an advertising executive who summarized the NCN experience: "They built a business and nobody came."[66] The reasons behind NCN's failure are multifaceted and represent broader issues newspaper companies faced when deciding what to do about the Internet.

In terms of content, newspaper companies highly valued the idea that NCN would provide a gateway to vast archives of content from newspapers across the country. The regional nature of the newspaper business in the U.S. made it difficult to create a national network where that did not naturally exist. There are a few newspapers such as the *Wall Street Journal, USA Today*, and the *New York Times* that have national distribution, but the daily newspaper business in the U.S. is largely based around metropolitan areas. NCN represented the attempt to address this market reality by allowing advertisers to use its national service to target advertising based on content from specific geographic regions.[67] Analysts thought this idea had merit, but by the time NCN's content feature was functioning, there was competition from "about 30 sites such as those run by Yahoo, Netscape, Microsoft and a dozen big papers."[68] The newspaper industry was praised for creating NCN as an early response to the Internet, but it was criticized heavily when it did not turn the vision into action quickly enough to ward off such competition.

The failure of NCN is also remembered as a missed financial opportunity for the newspaper industry given that it was created during the period when investors were lavishing new Internet ventures with capital and their initial public stock offerings were soaring in value. *BusinessWeek* reported that Cox Newspapers, one of NCN's owners, wanted to bring the venture capital firm Kleiner Perkins Caufield & Byers into the NCN ownership structure. Cox wanted the venture capital firm to invest in NCN and then assist the consortium with a public stock offering of its own.[69]

The proceeds from such a plan during these heady days would have endowed NCN with enough working capital to free its newspaper owners from further capital investment, while providing a significant return on their initial investment. Gannett remained neutral on the proposal, while Cox was supported by Tribune and Hearst. The other owners argued, however, that NCN should remain a private company. They believed the newspaper industry was flush with enough capital to fund the consortium and wanted its profits distributed back to the newspaper owners rather than public shareholders. The final vote defeating the outside investment strategy was 5–4, setting the stage for more disagreement over ownership issues when the financial performance did not happen as expected.[70]

Each of the nine founding newspaper companies had seeded NCN with $1 million, eventually investing more than $25 million total into the venture. However, NCN generated only $1 million in advertising revenue. More capital was needed to operate the business, but the New York Times Co. refused to invest, which forced the decision to close the business. The disagreements over how to capitalize NCN had resulted in deep rifts among the partners. Knight-Ridder's representative said the trouble began with the earlier rejection of the plan to accept venture capital and take the venture public during the overheated stock market of 1996. "The fallout from that plagued New Century Network," he said.[71]

Although newspaper industry leaders had done little to support the organization they created, the closing nevertheless was met with the rhetoric of failure similar to the sentiments expressed a decade earlier when Viewtron was shut down. An industry newsletter wrote:

> While the closure of the New Century Network was not completely unforeseen, the reaction of the newspaper business was. Sadness and disappointment reigned as word spread . . . that the on-line partnership of nine newspaper companies would end. The depth of feeling stemmed not only from a public failure, but also from a concern that the newspaper business had missed the boat—once again.[72]

The NCN closing was recognized as the industry's tacit admission that competing in the Internet era would be more difficult than originally assumed. It once again cast doubt that a traditional media industry could adapt existing business models to new market realities.

Most of the post-mortem analysis blamed the failure of NCN on its organizational structure. An executive with Hearst observing how NCN operated said: "Organizations of a number of co-equals can't turn on a dime."[73] The organizational structure was exacerbated by the diversity of the companies involved as expressed by a *Los Angeles Times* executive: "You had private companies and public companies and companies that were risk-averse and those that were risk-tolerant. You had big city papers and small chains. We share a need. But it was frustrating trying to come together."[74] This executive believed NCN problems reflected the independent nature of the newspaper business in the U.S. and was allowed to fail because "we didn't need it, and it was competitive."[75] Rather than invest in a common industry platform, most newspaper companies decided it was in their best interest to develop their own Internet operations. The businesses behind most newspapers decided that in the face of escalating value for Internet companies, it made more sense to own outright than to share. The question remained, however, if newspaper companies would invest to compete offensively in the emerging market or choose to react defensively.

Some executives viewed the failure of NCN as the industry's inability to grasp the larger ramifications of the Internet. A key executive in NCN's creation commented that its closure reflected that "Newspapers are reacting in very traditional ways to a very untraditional marketplace. And they're being superseded fast."[76] Nevertheless, there was a flurry of activity in the newspaper industry in the latter half of the 1990s as companies established websites and the production processes to support them. The history of how newspaper companies in the U.S. approached the web aside from the NCN consortium forms the remainder of this chapter.

Newspapers Independently Adopt the Internet

The period in which NCN failed also was a time of intense development as newspaper companies moved on their own to place content on the web. There was no clear pattern as to how newspaper companies approached the web during this period, but there was a lot of activity. The list of newspapers moving content to the web in the mid-1990s was a long one and included such early adopters as the *St. Petersburg Times*, *USA Today*, the *Philadelphia Inquirer*, the *Boston Globe*, and the *Arizona Republic*.[77]

Some newspaper companies that were part of media conglomerates joined forces with internal partners. Cox Newspapers, for example, was part of its owner's effort in 1996 to create a single Internet division to service the newspaper company, and its sister media companies, Cox Television, Cox Communications (cable television), and Cox Radio. Called Cox Interactive Media, this new unit was charged with creating a local Internet experience in all the markets in which Cox operated, including its newspaper markets. The flagship effort was based in Atlanta and attempted to combine the resources of Cox's media properties there, including the *Atlanta Journal-Constitution*, WSB-TV, and WSB Radio AM 750.[78] By early 1997, Cox had announced plans for 30 online sites to be operational by year-end, prompting CNET to call Cox's agenda one of the "most aggressive by any company to date."[79] Knight-Ridder created a network of its newspaper sites under the banner Real Cities,[80] and the *New York Times* emerged as an early Internet bellwether for the newspaper industry due to its national brand and distribution.

The *New York Times* debuted on the web January 22, 1996. The event was touted in a company press release as "part of a broad strategy to extend the newspaper's reach and create new editorial and business opportunities in electronic media."[81] The newspaper's publisher stated:

> Our site is designed to take full advantage of the evolving capabilities offered by the Internet. At the same time, we see our role on the Web as being similar to our traditional print role—to act as a thoughtful, unbiased filter and to provide to our customers information they need and can trust.[82]

The *Times* had tested the web the previous October when it posted content about the Pope's visit to New York.[83] The experiment drew criticism from financial executives of the newspaper who viewed it as an editorial exercise without any thought given about how to pay for it. An executive recalled the episode:

> . . . it was really an exercise of the publishing system more than anything. . . . The thing went up and we were very happy with it. And I got a call from the fellow who was running the company at the time, screaming at me about how it didn't have any ads in it.[84]

The incident is illustrative of many early endeavors when newspaper companies were more concerned about a content presence on the Internet than how to establish an advertising business model. When the *New York Times* formally launched its web presence, however, the advertising issue was addressed. A press release said the newspaper intended its website to become a money-making business, noting that "charter advertisers" included Toyota and Chemical Bank.[85] The *New York Times* understood its national brand would allow the newspaper to compete online more broadly than other newspapers, given that its printed newspaper was distributed nationally. Two years after its launch, the national market approach appeared to be working. The paper's website had achieved four million registered users by May 1998, with 83 percent of those registrants living outside metropolitan New York.[86] The national—even global—reach of the *New York Times* allowed it to pursue a different Internet strategy than regional newspapers.

The *Boston Globe*, by contrast, provides an example of a newspaper that followed a regional approach. It created an electronic publishing division that pursued partnerships with other Boston area media to launch a regional Internet destination. The website was named boston.com and was marketed as "New England's first Megawebsite, a one-stop interactive resource on the Internet's World Wide Web." The newspaper's content was featured alongside material from more than a dozen Boston television stations, radio stations, magazines, and other sources of content.[87]

The *Boston Globe*'s approach was considered a radical departure for the newspaper industry. A trade publication described boston.com as "not what you'd call a newspaper on the Web," adding that it "breaks traditional rules of branding by partnering with broadcast and print competitors." The article described the effort as an attempt to create "a cyberbrand with greater clout—a whole greater than the sum of its parts" by featuring content as the drawing card.[88] The article illustrated that much of the newspaper industry held a narrow view of what an online newspaper should be, but companies continued to push the boundaries. The *Washington Post*, for example, partnered with competing media—*Newsweek* and ABC—to launch ElectionLine, a website focused on the 1996 elections.[89]

As noted earlier, Boczkowski wrote that years of online experimentation by the newspaper industry culminated in 1995 when most newspaper companies adopted the Internet and the World Wide Web as the electronic delivery platform. However, deciding on the web as a delivery platform did not mean the newspaper industry's online evolution had reached its conclusion. The variety of partnerships and types of websites newspaper companies developed in the early years of the Internet demonstrated that an acceptable electronic publishing model had yet to be achieved.

The latter half of the 1990s was a transitional period for newspaper companies and their online efforts. Some newspapers hedged their bets by investing in a web presence while also retaining other partnerships. The *Houston Chronicle*, for example, established a website, but continued to post content through its Prodigy affiliation and agreed to market Prodigy as the "preferred means" of accessing the Internet.[90] Over time such alliances with proprietary services would give way to exclusive web-based operations, but the industry's experimentation process was far from over. The following sections examine several pertinent business issues newspaper publishers and editors faced when deciding on how to approach content, access, and financial models.

Newspaper Business Versus Information Business

Picard wrote that a business model reflects "the conception of how the business operates," including "its underlying foundations, and the exchange activities and financial flows upon which it can be successful."[91] By this definition, most newspaper executives in the late 1990s did not view the business model of an online newspaper as being very different from the business model of a printed newspaper. In both cases, the fundamental objective was to produce content that attracted an audience of sufficient size that would, in turn, attract advertisers.

The emergence of the web, however, forced the newspaper industry to assess the fundamental nature of its business. The American Society of Newspaper Editors (ASNE) framed the issue in an essay that asked: "Are you an organization that supplies newspapers or are you an organization that supplies information?"[92] In proposing to explain why the industry's response was so important, the essayist wrote: "Remington and Underwood saw themselves as being in the typewriter business. IBM saw itself as being in the word-processing business. The rest is history."[93] In a similar vein, a speaker at an industry conference said newspaper companies should "be prepared to decouple from paper," adding that "perhaps when

I come back a few years from now, it'll be the News Association of America,"[94] rather than the Newspaper Association of America.

Whether or not newspaper companies defined their product as news or information rather than by the physical product, a newspaper was much more than a philosophical issue. How newspaper companies defined their business would determine how they operated. In retrospect, the production process of a daily newspaper led to many preconceived notions that influenced decisions about producing content in the online arena. Lapham recognized the difficult challenge newspapers faced in this regard. "Reinventing itself is a tall order for an industry that works under constant deadline to produce a new product each day," she wrote.[95] Newspaper companies had invested in computer-based production systems in the 1980s and believed they should leverage those systems to also produce content for the Internet. A Knight-Ridder executive said of his operation: "we figured out a way to automate" the transfer of content from the printed newspaper to the online version "so that it didn't cost a lot."[96] While efficient, the approach allowed the printed product to control the flow of information and reduced the online edition to ancillary status.

Cultural Artifacts Take Precedent

The process of transferring content from a printed newspaper to its online sibling became a ubiquitous practice in the newspaper industry. So much so that the software applications developed to automate the process became known as "shovelware."[97] This example illustrates how business processes applied to the web echoed the cultural artifacts of the printed products and often became an obstacle that prevented newspaper websites from absorbing and reflecting what was happening elsewhere in Internet-based media.

Chyi and Sylvie described the production processes that relied on shovelware as "moving whatever is published in the newspaper onto the Web without further developing the information," adding that it "requires a very small staff."[98] Keeping labor costs under control within their online versions was indeed a goal of newspaper managers who chose to deploy shovelware. In the early going, however, the motivation was less about expense control and more about leveraging what was perceived to be the most valuable asset a newspaper company owned—its content.

Anticipated Value of Content Influences Action

A prevailing belief among newspaper publishers and editors at the outset of the Internet was that newspaper content was valuable and represented a competitive barrier to entry given the large staff of reporters, editors, and photographers required to create it. The newspaper industry believed that placing its valuable content online would make it more accessible and, in turn, even more valuable. This view influenced the efforts at many newspapers to move historical content into online archives, allowing consumers to access newspaper morgues—an industry term for libraries—for the first time. In many newspapers, the effort to create digital archives and online Internet versions became a co-mingled process.

How newspaper companies valued their content at this time is illustrated in an example from the industry's trade press. A special report about online archives opened with a reference to the 1960s television sitcom *The Beverly Hillbillies*. In the opening sequence of the show, the patriarch Jed discovered oil, which the theme song describes as "black gold."[99] The article made the connection as follows:

> Much like the pre-Beverly Hillbilly Jed, North American newspaper publishers sit on pools of wealth. . . . The black gold is information, potentially gigabytes of it, archived in digital form. Intelligence scribed by a room of smart, experienced reporters and editors. Images captured by talented photojournalists. Graphics created by savvy info-design specialists. Not to mention the display ad for Bernie's Grocery or the classified for a cherry 1957 Chevy.[100]

Through the Internet, newspaper executives saw new possibilities to extract value from their troves of data. As one executive expressed: "Publishers can put material out for the whole world. . . . We'll see the death of the middlemen."[101] Newspaper companies had relied on deals with commercial database companies to compile their material for sale in the secondary research market, often receiving only token royalties.

The lure of extracting more value from archived material and the relative simplicity of shovelware systems contributed to the rapid increase in the number of newspapers migrating to the web in the late 1990s. However, newspaper companies were more fixated on the process of transferring content than they were with what to do with their online content. In many cases, the result was a static presentation of the printed paper in electronic form, sparking a wave of criticism.

The Backlash against Shovelware

A leading industry executive said one of the biggest lessons newspaper companies had learned from their earlier online efforts with videotext and propriety services was the need to provide consumers with an engaging experience. "Don't just replicate the newspaper. . . . Text on screen isn't enough," this executive stated.[102] Yet that is precisely what most newspaper companies did with the Internet in the 1990s. Online editions were electronic renditions of the printed newspaper that reflected little of the functionality necessary for making the content interactive. A more detailed discussion of interactivity as it relates to this period is included later in this chapter. Meanwhile, the discussion of online production processes illustrates that newspaper companies did not convince their critics they were heeding lessons learned from previous online experiences.

As one commentator asked and answered: "What will make you stand out on the Internet? Certainly not dumping the entire contents of your newspaper on the Web." This criticism said newspaper companies had "to learn how to select material and package it in exciting ways," adding that success depended on treating online content as more than an archive.[103] Chyi and Sylvie found that reliance on shovelware resulted in online editions that failed to differentiate from the printed newspaper. They wrote: "Due to the almost inevitable relationship with—or dependence upon—their print counterparts, many electronic newspapers have not become an independent entity, not even conceptually."[104]

Most online newspapers relied on repurposed printed material for the majority of their content because it was an efficient production process, but that "does not necessarily imply quality content," according to Chyi and Sylvie, who referred to the tactic as "folly."[105] They concluded that the practice undermined the newspaper industry's ability to compete, stating that "online newspapers that lack original content do not appear to be engaging in successfully strategic behavior." This lack of original content made it difficult to establish an independent brand identity for the online products, which they argued was necessary for long-term viability. "Newspapers are relying on the brand preference established by their print versions, hoping that the print newspaper's credibility and desirability will transfer to the new medium," they wrote.[106] Despite such observations, newspaper companies were unapologetic for attempting to build their online presence based on the brand recognition of their printed newspapers.

The publisher of the *New York Times,* for example, stated that one of his key strategies was "to take the brand we have today and to translate it for this new medium." He acknowledged, however, that the online product would likely turn into something different over time. "Some of the parts will be shockingly familiar to all of us. Twenty and twenty-five years from now, other parts none of us can even imagine," he said.[107] The use of such a long-time horizon, however, illustrated the newspaper industry's belief that there was ample time to sort out operating issues. Newspaper companies seemed to operate their early websites on the premise that short-term decisions would have little effect on long-term strategic concerns. The most illustrative example of this notion was the practice of embargoing news coverage, preventing publication on a website until after the material had appeared in print.

Online Embargos Echo Culture of Control

Publishers and editors considered their printed newspapers preeminent over their online versions during the early days of the Internet. This favoritism was not only reflected in the repurposing of content, but also with the timing in which the repurposing occurred. In most cases, newspapers embargoed their own content until after it had appeared in print. The *New York Times* summarized such practices:

> Virtually all of the hundreds of Web sites set up by the nation's newspapers still do not take full advantage of one of the Internet's most compelling features: its immediacy. Instead, most newspapers, including the *Washington Post*, the *Los Angeles Times* and the *New York Times*, rely on reports from news agencies to update their sites during the day, while holding back their crown jewels—what their own journalists have discovered—until the actual papers are nearly on the street.[108]

Newspaper editors explained the rationale for content embargoes as an attempt to preserve the role of the printed product. The practice also revealed how newspaper journalists were deeply rooted in competitive traditions.

Journalists did not want material available that would alert local television and radio stations, and other newspapers in some cases, about exclusive material. Therefore, allowing their work to be read in advance of print publication was a concept many editors and reporters could not fathom. An editor involved in online operations stated: "The whole idea of scooping ourselves is troubling to a lot of people. There are grave concerns within the newspaper industry about the extent to which new media are going to cannibalize the existing services that we provide to our consumers."[109] As late as 2000, an industry survey found that 45 percent of its print respondents said they did not allow their websites to scoop their printed newspaper. While this represented a decline from 58 percent in the prior year, it still reflected widespread use of embargoes.[110] The embargo issue underscored how complex the newspaper industry's relationship with the web was during this period. The newspaper industry's reliance on content repurposing reflected the notion of the web as an extension of the printed product, but at the same time, content embargos reflected the fears of the web as an emerging competitor.

The practice of using wire service material rather than original content to update online newspapers between print editions also may have contributed to consumers undervaluing the newspaper's own content. Researchers, including Leckner[111] and Lee,[112] discussed a phenomenon described as the commoditization of news brought on by the Internet's capability of aggregating large amounts of information from multiple sources and making it readily available to anyone who wants to receive it. As Lee stated: "The Internet . . . created a platform for easy distribution of news information."[113] By embargoing their own material and exposing readers to the extensive amount of generic information from services such as the Associated Press and Reuters, the newspaper industry perhaps contributed to the commoditization process, especially in the areas of national news, financial data, and sports coverage.

On a more fundamental level, however, newspaper companies were being challenged to assess their position in the flow of information. A media critic wrote:

> Newspapers might begin to think about reversing their long-standing priorities, recognizing that everyone with electricity has access to more breaking news than they provide, faster than they provide it. They should, at last, accept that there is little of significance that they get to tell us for the first time.[114]

The Internet represented another reminder that newspaper control over information had been declining in the face of electronic competition. Editors had believed that newspapers succeeded in the face of electronic competition because they provided more depth and perspective than television or radio. The Internet, however, challenged those assumptions with limitless storage, the capability of linking vast amounts of related material, and by providing access to a diversity of sources.

Media critics said newspapers would be forced to change and, as one wrote, acknowledge that "they don't get to tell us only what they think we should know," adding that "they'll also have to really listen to us, not just pretend."[115] This reflected the notion that newspapers could alter their relationship with their readers by using technology to foster a dialogue. The Internet was seen as an interactive medium, but the newspaper industry's early websites were widely criticized for not taking advantage of this capability. The following section examines the calls for newspaper websites to become interactive and how newspaper companies responded.

A Call for Interactivity

Interactivity was discussed in the Introduction to provide an understanding of its meaning in a broad conceptual context. In revisiting the topic as it applies to the Internet and newspaper websites, interactivity is explored in this section as a pragmatic attribute. Newspaper companies were expected to use interactive functionality to create a two-way communication system and establish dialogue with their constituencies. An executive who pioneered the use of the Internet in the newspaper industry wrote:

> Effectively used, an Internet presence can transform static information, whether articles or ads, into real-time interaction—between sources and the newsroom, between the newsroom and readers, and between readers and advertisers.[116]

Just because the Internet could facilitate interaction, however, did not translate into why newspapers should want to solicit interaction. Therefore, it makes sense to explore how this concept was considered at a practical level in the early days of the Internet and how it affected what newspaper companies did.

The executive quoted above said interaction would allow newspapers "to cement relationships with customers."[117] Some called on the newspaper industry to embrace interactivity as a natural way to offset the effects of news commoditization and the diminished role of reporting on events as they happen. Hollander supported the notion that interactivity would foster community and allow newspapers to "recapture from talk radio the position of community forum." He maintained that "people can easily find information if they want it," but indicated that newspapers could differentiate themselves online by creating "the place where the community communicates with itself, where ideas are debated, policy is discussed, problems are confronted."[118] Lapham also wrote of newspapers using online tools to become "a facilitator of public discussion."[119] Therefore, the Internet was seen as providing the tools newspapers could use to elevate their role far beyond soliciting feedback through letters to the editor.

It represented, as Pavlik noted, a dramatically altered content paradigm—one that had been at the core of newspaper content for decades.[120] He wrote: "No longer can most journalists and editors be content merely to publish the news. Instead the process is becoming much more of a dialog between the press and the public."[121] But Pavlik recognized that "not all news organizations are comfortable with this emerging model" because they feared that interactivity would require sharing control of the message.[122] Pavlik observed that a newspaper's centralized control over its content gave it immense power in a particular market. Over time, editors came to understand that such power allowed them to set the agenda for their market, a role they viewed as a key responsibility rooted in journalism history.[123] Real-time feedback questioning news coverage decisions and challenging editorial authority was not an appealing proposition for most newsrooms.

However, if newspaper editors were uncomfortable supporting interaction of this nature, then newspapers were left to determine how they would take advantage of the Internet in other ways. Pavlik addressed this quandary in a rhetorical question directed at the industry, in which he asked if the Internet "is little more than another delivery system for 'old media'—even if it's a potentially better delivery system—what's all the fuss about?"[124] Many of the newspaper industry's early online efforts did not represent an answer to this question. They were content with using the Internet as a distribution system. But over time, the industry explored other aspects of Internet functionality and experimented with interactivity—not principally as a mechanism for dialogue and feedback, but for content selection and presentation.

Linking content with interactivity led to the construct of personalization, a term that described functionality for allowing users to select the information they wanted from a menu of choices.[125] In the newspaper industry, personalization had been expressed as "The Daily Me,"[126] which held particular fascination with editors. In this model, editors retained significant control given that content remained centrally produced. User interaction was limited to selecting the information they received, and in some cases, determining how the information was presented on the screen.

This model represented a step forward for newspapers in deploying Internet functionality and was adopted widely. Some newspapers—the *Christian Science Monitor* and the *Wall Street Journal* are two examples—thought of personalization as such an important functionality that it initially was sold to readers as a premium service.[127] Critics cautioned that although personalization starts with the user's selection process, it was only partially interactive because it relies on "computerized 'butlers' or 'agents'" that "are acting on the reader's behalf," to process the information that is presented.[128] As such, personalization became a tool of the Internet, but it was not a construct that defined the content relationship between an online publication and its audience.

Kilker observed that personalization providing users with basic levels of control did not go far enough in using the functionality of the Internet and would not satisfy consumers who have "additional expectations." Kilker concluded that "users expect to control their media consumption through not only filtering, but also time-shifting, archiving, and reformulating content."[129] As the Internet unleashed increasing expectations, consumers expected traditional media companies to keep up with developments. Therefore, newspaper publishers were facing a marketplace with expectations very different from the one traditional media had served successfully for decades. These differences would affect publishers' ability to transpose familiar business practices onto an online market that had spawned new competitors free of the newspaper industry conventions.

New Competition

Newspaper companies in the latter half of the 1990s faced an unsettled online environment. Proprietary online systems such as AOL and Prodigy still were adding subscribers, and Microsoft had launched MSN, a proprietary system of its own. As discussed in the previous chapter, these proprietary systems had been allies of the newspaper industry. Even Microsoft had forged partnerships with several newspapers, including *USA Today* and the *New York Times*.[130] However, Microsoft's decision to target local markets with an extension of MSN called Sidewalk drew the newspaper industry's ire. A trade article referred to Microsoft's founder Bill Gates as the industry's "favorite devil in a blue suit."[131] When trade articles reported that Microsoft planned to spend as much as $600 million developing online content—more than the entire newspaper industry would invest in online projects—publishers realized the investment needed to compete had been elevated to unexpected levels.[132]

Newspaper executives began to understand that online platforms provided friends and adversaries new ways to compete. As the meaning of online morphed into meaning only the Internet, the speed at which it developed also led to new competition eager to exploit what many saw as the newspaper industry's myopic view of the media landscape. An executive with Yahoo, one of the emerging competitors, observed that newspapers have "killer, killer content, but too much pride in their own distribution."[133] The implication of that remark was that newspaper companies were too enamored with their own way of operating. The result was a slow response to market changes shaped by new competition.

Only two years into the Internet era, a trade association report identified "as many as 40" new online content companies and several more online classified advertising businesses that could be deemed as competitive to newspapers.[134] A technology writer with the *New York Times* wrote the newspaper industry found that by taking "its revenue model online, it becomes vulnerable to a phalanx of new competitors," because "geography no longer matters."[135] An industry analyst later described the result of this activity when he wrote that "the components of what we historically know as the newspaper have become unbundled."[136] In considering these effects of the new online competition, this chapter next examines the portal phenomena followed by a section exploring the emergence of online classified companies.

Portals Threaten Using Familiar Model

Newspapers were accustomed to competing in local markets against other print publications, television, and radio. The Internet, however, led to national competitors with plans to siphon away advertising from local markets as well. As these companies quickly evolved from their roots as search engines or content aggregators, they became known as portals. These portals represented a significant early effort by non-media companies—for the most part—to apply media-like business models to the Internet. The business objective was to aggregate an audience through a variety of services and then monetize that audience through advertising supplemented with subscriptions and other transactional revenue.

According to Hinton, the term "portal" surfaced in industry and trade sources in late 1997 and became a mainstream topic for discussion in early 1998.[137] Shaw defined a portal as a website with "at least five core features," which were listed as search, news, reference tools, online shopping, and communications (such as e-mail, chat, or message boards).[138] The portal construct emerged out of a design "to perform two functions." The first function was to provide utility through the types of features just listed, while the second was "to concentrate users and create large audiences, thereby generating revenue, typically through advertising."[139] Despite business models that relied on advertising support, early portals were not described by analysts as media companies. Traditional media and telecommunications companies were not involved in creating the portal phenomenon, and as Mansell observed, they also did not "move aggressively into the portal market despite the market potential and their existing strengths."[140] In effect, portals were seen as developing independent of traditional media, which did little to mount a competitive resistance.

AOL and MSN launched portals on the Internet using their roots as proprietary online services. But the portal companies attracting the most attention were those that evolved from Internet search engines such as Yahoo and Excite. As the market developed, traditional media—especially newspaper companies—were seen as in danger of being outflanked by portal competitors that were not so much defining a new market as they were redefining an old one.

For example, publishers and editors maintained that newspapers would be best at translating to the Internet their ability to filter and organize content. However, Shaw noted that organizing content was a major strong suit of the portal companies:

> Portals have also taken on the equally important role of aggregators of Web-based content, a role that many Web surfers wholeheartedly welcome. A good number of Internet users believe portals are helping to tame the Internet by organizing its abundant resources and helping users better focus their surfing time.[141]

The acceptance of this functionality by consumers further enhanced the portals' media-type business model and allowed them to occupy a position many newspaper publishers thought would be theirs.

Damsgaard said the successful portals went beyond content organization and also captured audience by using their own search engines and proprietary e-mail and messaging platforms to make their websites destinations on the Internet rather than portals to outside content. Damsgaard described this as "closed self-referring systems or wall-gardens."[142] Mansell referred to these efforts as "supplier monopolization strategies" and credited many of the portal companies with achieving a perception of exclusivity or product scarcity through the "balkanization of service offerings."[143] It was a market achievement newspaper companies had originally sought, but failed to achieve, through their partnerships with the proprietary online services.

The success of the portal companies and the business models that supported them reflected a market that had come full circle in a short time. Consider Hinton's conclusion:

> By establishing centralised points of access—the construction of transactional spaces to mediate users' interaction with the medium—portal sites effectively re-establish the structure of power and control that early proponents of the Internet hoped would be rendered useless and archaic in the distributed, anarchic medium of the Internet.[144]

In essence, the portal companies established a media business model on the Internet at the expense of newspaper companies and other traditional media companies. As the chief executive officer of Yahoo stated: "I don't think old media is what people are going to spend most of their time doing on the Internet. This paradigm needs its own inventions, its own methods, its own way to go forward."[145] Traditional media had been too slow to react, giving credence to the description of these upstart companies as the "new" media.

Meisel and Sullivan were among the researchers to study and write about portal companies as new forms of media companies. They wrote that portals became media in that they deployed familiar media business models, but were "new" in that they operated on the Internet and were not tethered to any traditional medium. These "new media" companies, they wrote, emerged from technology innovation, but also were often collections of smaller businesses brought together through mergers and acquisitions. But Meisel and Sullivan also observed that the emergence of these portal companies was aided, at least in part, by the mistakes of competitors.[146] Even if the newspaper industry was complicit in allowing the emergence of new media competition, publishers had to confront the new reality in moving forward.

The Internet had reduced barriers to entry (large capital outlays, economies of scale) that had protected newspapers from many forms of direct competition in the past.[147] Not only did the reduced barrier to entry facilitate the rise of news competition, it also contributed to the development of businesses taking aim at niche markets. These new competitors were focusing on "consumer decision" content that included such things as employment, real estate and automotive classified advertising, and entertainment listings, which were all areas "where new technology can replace static newspaper lists with searchable online-city guides, intelligent search agents and personal content."[148] The emergence of this type of competition was most threatening to newspaper classified advertising. The following section explores how the newspaper industry reacted as it faced new electronic threats to its very important classified franchise.

The Classified Advertising Threat

In 1990, newspapers derived $11.5 billion, or 35 percent of overall advertising revenue, from classified advertising. By 2000, classified advertising had grown to $19.6 billion and represented 40 percent of total newspaper advertising—and an even greater percentage at some large metropolitan dailies.[149] In large measure, however, the growth reflected the industry's practice of raising rates to help offset declines in volume. In the first half of the 1990s, for example, classified advertising rates "rose an average of nearly 8 percent per year," outpacing the inflation rate of 3 percent annually.[150] An industry analyst commented that newspaper publishers "worry about preserving profits, and they ought to be worrying about preserving [the] franchise."[151] The remark suggested the rate increases had allowed newspaper companies to maintain revenue growth, but had obscured erosion of the industry's share of classified advertising.

Therefore, within the context of this book, it is important to understand that the arrival of Internet-based competition for classified advertising occurred when newspapers already were struggling with this category. By 1996 apartment guides printed by competitors, for example, were claiming $1 billion in annual revenue that at one time had been in newspaper classifieds.[152] A study by an industry trade association found that employers were spending about 80 percent of their recruitment budgets in media other than newspapers, while about 58 percent of real estate classified advertising was spent elsewhere. The study prompted the association's economist to warn: "The entire classified category should be considered at risk." The economic importance of classified advertising was illustrated by an industry report that showed a newspaper with a typical operating margin of 14 percent would see that margin decline to 9 percent with a 25 percent decline in classified advertising revenue. Should classified revenue fall by 50 percent, the study said operating margins would decline to 3 percent.[153]

Many newspaper companies believed early on that the Internet would be useful in mitigating their risk in classified advertising rather than spawn outright competition. As such, newspaper companies often provided online listings for free or at a reduced price as part of the print-based purchase.[154] In doing so, however, newspapers devalued their online classifieds and contributed to a market climate where low-cost competitors could thrive. Newspapers had learned from the popularity of "buy-and-sell sites" on electronic bulletin boards and early videotext systems that classified advertising likely would perform well on the Internet.[155] To its credit, the industry launched several classified advertising ventures and most online newspapers featured classified advertising sections. But the industry was criticized for failing to implement an effective online classified strategy. In terms of production, newspaper companies treated classified advertising in much the same way they treated news and information content. Automated systems were developed to transmit classified advertising from the printed newspaper to the online newspaper.

Although newspaper companies could claim their online editions featured classified advertising, the presentation was often static, providing users limited options for searching and interacting with the advertising. A trade publication summarized the newspaper industry's approach to online classified advertising:

> . . . although many newspapers have set up Web sites that include classified advertising sections, the strategies and functions of those sections are generally weak and ineffectual in comparison to the innovative systems deployed by the leading non-newspaper cyber-classifieds companies.[156]

Mings and White cited other industry sources that also concluded newspaper companies were outmaneuvered by new Internet companies, including monster. com, hotjobs.com, and e-Bay.com in areas such as "immediacy; customizability; and special functions such as keyword search and retrieval."[157] A research company specializing in Internet-based businesses predicted in 1999 that newspapers would lose more than $3 billion in classified advertising revenue over five years to Internet competitors.[158] Another research firm wrote that "newspapers are going to suffer in that they're going to find an erosion in the big three classified categories—jobs, homes and automobiles—and that's going to put pressure on their traditional business."[159] As the scope of the newspaper industry's classified problem emerged, critics began expressing the issue in terms of newspapers losing the classified franchise.

For example, an industry consultant stated: "It is not going to be a matter of if they lose their classified lineage, but when and to whom."[160] One of the more critical essays accused newspaper publishers of forfeiting their classified advertising business without a fight:

> This loss of franchise is not the result of pitched battles between the traditional holders, the newspapers, and the new upstarts. Rather, it's all happening very peacefully as the newspapers simply watch and study events from the sidelines. The upstarts aren't "winning" market positions that someone else held online, rather, they are simply filling in a void that the newspapers as an industry have seemingly refused to value.[161]

Industry leaders acknowledged some of the criticism directed their way. For example, a trade association's technology executive conceded that newspaper classifieds should be "more facile to work with," indicating that improvements were needed in search and other interactive functionality.[162] As for the overall strategic approach, however, industry executives continued to position their online classified advertising as complementary, not competitive, to their printed product.

An example of this approach was Classified Ventures, a partnership that included Times Mirror Co., Tribune Co., and the Washington Post Co. The venture launched online classified advertising sites, including cars.com and apartments.com, and structured distribution agreements through newspaper company websites. But as late as 1997, the head of the venture stated that over the next three-year horizon, "we believe the Internet is not going to have much effect on the newspaper business." Based on that outlook, he added, "we don't want to do anything in any way to threaten the traditional print business."[163] The statement about timing reflected the newspaper industry's attitude about technology—that is was something that would have greater impact in the future and that there would be time to figure it out. By not giving their own online classified businesses free reign in the market in the meantime, however, newspaper companies made themselves even more vulnerable to outside competitors.

An industry commentator wrote that classified advertising had been the newspaper industry's most profitable source of revenue, referring to it as the industry's "secret weapon."[164] In the latter half of the 1990s, however, evidence mounted that erosion in the classified advertising segment was contributing to an overall slowing of total newspaper advertising growth. When the head of the industry's trade association attributed some of the decline to the effects of Internet competition, the commentator wrote in 1999 "this may be the beginning of the end."[165] In dollar terms, he was prescient. Total advertising revenue, including classifieds, peaked at U.S. newspapers in 2000 at $48.7 billion.[166] Nevertheless, newspaper companies remained vigilant about their overall profitability, and maintaining profit margins was a key business driver in 2000. The final section of this chapter examines the newspaper industry's financial concerns and the influence they had on the industry's Internet strategy.

Profitability Versus Viability

The newspaper industry was beginning to experience the new market reality during the 1990s, even though many of its executives did not fully appreciate at the time that long-term patterns were forming because they were somewhat masked by financial growth trends. For example, the industry's overall advertising revenue grew 43 percent from 1990 to 1999. However, this nominal expansion was not enough to keep pace with total advertising growth in the U.S., and newspapers lost market share to other media.

Nevertheless, newspapers were businesses in demand. Lacy and Martin reported that 856 U.S. daily newspapers were sold during the 1990s, including 153 newspapers "that were sold more than once." This level of activity meant that "sales of daily newspapers in the 1990s were higher than sales for the previous two decades," the researchers wrote.[167] The merger and acquisition activity underscored the view that newspapers during this period remained favorable businesses to own because they were generating significant cash flows and operating profits.[168] At the end of the decade, newspaper companies averaged a 20 percent pre-tax operating profit, compared to 13 percent for all U.S. industries.[169] The Tribune Co.'s deal to buy Times Mirror Co. for 100 percent above its stock price demonstrated the industry's conviction that newspaper companies held long-term value.[170]

However, the erosion of advertising market share began to take a toll on the industry's collective psyche in the latter half of the decade when newspaper companies lost their coveted advertising leadership position. Consider the 1996 industry report from the trade association's chief economist:

> Other media have been gnawing away at our dominance for a generation. Our declining performance was not a compelling issue before the last recession because all-media advertising growth continued to outpace general economic growth. This share erosion, though, became painfully apparent during the advertising recession. . . . Nothing in my forecast suggests a reversal of this trend. The best I can expect is a slower loss in our share. The long-term implications are dramatic. Our claim as the No. 1 advertising medium may soon be history.[171]

By the following year, this forecast had come true. Television—bolstered by the expansion of cable networks—surpassed newspapers in 1997 as the largest advertising medium. Television edged ahead that year with a 23.8 percent share of the market compared to newspapers' 22.2 percent market share.[172] For the first time in decades, the newspaper industry was forced to respond to financial numbers that indicated tangible changes in the media landscape.

In confronting their new reality, newspaper publishers understood the Internet needed to be part of a comprehensive strategy, but the uncertainty about what to do and how to do it was unnerving. The difficulty in deciding an Internet strategy was exacerbated by financial concerns. As Boczkowski wrote, "profitability was a particularly sensitive issue for online newspapers."[173] For example, several prominent newspaper companies in 1998 reported sizeable losses from their online newspapers. Knight-Ridder estimated online losses at $23 million, while Tribune lost $35 million. The New York Times Co.'s loss was estimated at between $10 million and $15 million, and Times Mirror's loss was estimated at $20 million.[174]

The losses associated with Internet newspapers reminded many executives of their experiences with videotext projects a decade earlier and led to introspection about the newspaper industry's ability to effect a different result. One commentator summarized the prevailing sentiment:

> . . . it appears that many people with financial responsibility for online news operations . . . are weary of hemorrhaging cash. After far too many years of talk about revenue and nary a peep about profit, they would very much like for someone,

anyone, to show them the money . . . the questions that one would assume are most critical to news organizations online . . . . "Where is the money?" and "How long will we have to wait to see it?"—should have been answered long before now.[175]

The industry's frustration also resulted from years of experiments showing not where the money was, but where the money was not. For example, newspaper companies were generating a considerable amount of revenue for their printed papers from subscriptions. The industry averaged about 25 percent of total revenue from subscriptions during this period.[176] In the early years of the Internet, however, newspaper companies learned they could not depend on subscriptions to provide any meaningful online revenue.

One group of researchers observed that "as far as subscription is concerned, the situation is not optimistic for online papers. The Internet culture is characterized by free information, and very few online papers have actually started charging readers. . . . Those that tried to charge readers typically experienced a sharp drop in readership."[177] The researchers acknowledged that publishers of large national newspapers such as the *Wall Street Journal* would find a portion of their audience willing to pay for an online subscription, but concluded that "for local dailies, charging readers is even less realistic as a source of revenue."[178] Lee wrote that newspapers found their readers were unwilling to pay for content. "Worse, the charging scares away the readers," he added.[179]

Some newspaper companies attempted to offset the inability to sell subscriptions to their online newspaper by selling access to the Internet itself.[180] These newspaper companies established their own Internet Service Provider (ISP) operations and attempted to compete directly with telecommunications providers in their markets. In other cases, they partnered with telecommunications providers to offer either a co-branded or private-labeled service. One prominent example comes from Landmark Communications Inc. and Knight-Ridder Inc. The two newspaper companies created a partnership known as InfiNet, which provided ISP services to its owners' newspapers and also sold services to other newspaper companies.[181]

The industry's trade association did not endorse newspaper companies' foray into the ISP business. Instead, it urged newspaper companies to proceed with caution:

> . . . it's only a matter of time before the profits from providing Internet access grow razor thin. Therefore, consider access as simply an entry strategy to leverage the cost of the on-line business and to forestall [competitors]. Remember: what customers really buy isn't the pipe at all. It's access to community, conversation and culture. Readers want to be where the pipe leads.[182]

The industry group was prophetic in its forecasts about the ISP business, and in a historical context, the newspaper industry's brief flirtation with it can be summarized as inconsequential.

With paid subscriptions ruled out as a short-term revenue source, early online newspapers experimented with all types of advertising. In addition to classified advertising, online newspapers deployed banner advertising and sold sponsorships for specific sections of a website or email delivery of a website's content.[183] As the Internet expanded, however, researchers warned that online marketing would allow advertisers to bypass newspapers and other traditional media. McMillan wrote that "advertisers have new ways of communicating directly with target audiences."[184] To counteract this effect, an advertising executive with the *Los Angeles Times* said newspapers should use the Internet as "a logical product expansion . . . to create a pipeline between consumers and advertisers."[185] To remain relevant in the online arena, therefore, traditional media companies—including newspapers—believed they would have to deliver what they always had: audience.

By 1999, online advertising had reached approximately $2.8 billion, which was about 1.3 percent of total advertising spending. The online newspaper share of the total was miniscule, however. And as the aforementioned losses attest, selling enough advertising to offset the cost of operations proved to be problematic. The industry's combined revenue from online newspapers was so inconsequential that, the trade association did not report the number until it passed the $1.2 billion mark in 2003.[186] By then, overall spending for online advertising was $7.2 billion.[187] The newspaper industry had some online advertising success, but the verdict on a long-term value proposition was unsettled in the early 2000s.

Boczkowski suggested that the newspaper industry had collectively reacted to its circumstances, as would be expected of a mature industry—treating the changes in the marketplace as evolutionary and making slow and methodical adjustments. He wrote: "Newspapers have neither stood still in the midst of major technological changes, nor incorporated them from a blank slate, but appropriated novel capabilities . . . from the starting point of print's culture."[188] The late 1990s for the newspaper industry was as much about confronting its own culture as it was about confronting new competition. As Boczkowski wrote, its "actors have attempted to create a 'new' entity preserving the 'old' one."[189] He explained:

> That is, they have tried to transform a delivery vehicle that has remained unaltered for centuries, and whose permanence has anchored a complex ecology of information symbols, artifacts, and practices, while simultaneously aiming to leave the core of what they do, and are, untouched.[190]

Even though the ties to print culture were strong, newspaper companies understood they had lost ground to their old nemesis, television, and attempts to exploit the Internet as a media platform were falling behind new competition. Newspaper executives faced a muddled concept of what their market should look like, but knew that audience consumption of news and advertising was changing and would continue to change at a rapid pace. It was a difficult period for those operating in a tradition-bound culture to grasp the requirements of change.

Nevertheless, many in the industry set aside doubts about the financial viability of online operations and continued to invest in them. For example, the Washington Post Co. in 1998 announced plans for $100 million in new Internet spending, while the New York Times Co. told its investors to expect losses from online operations to grow as the company continued to spend on development.[191] From an overall industry perspective, however, no coherent strategy emerged for exploiting online technologies. Readers defected in larger numbers, advertisers sought better solutions, and the investment community wanted answers from newspaper companies about how they planned to fix their problems. The cumulative effect of these issues will be addressed in Chapter 6.

Discussion

The newspaper industry during the 1990s had its critics: those who believed publishers were complicit in enabling the rise of new media companies by failing to mount an aggressive competitive response. These critics argued that newspaper companies erred in choosing not to invest in innovation at levels required to fend off competition from the portal companies or the new purveyors of classified advertising services.

Clearly, the newspaper industry was surprised by the speed at which the Internet became a media platform for consumers. As this chapter recalled, newspaper companies were just getting comfortable in their partnerships with proprietary online services when the Internet emerged. Once the Internet phenomenon was understood, however, newspapers were collectively interested in exploiting it. The operative word is "collectively," given that the newspaper industry's most aggressive action during this period led to the creation of NCN, a consortium intended to provide the newspaper industry with a competitive Internet platform.

The case study of NCN serves as an analogy for the broader perspective of the newspaper industry at the beginning of the Internet era. NCN represented the grandiose dreams of an industry that proved incapable of executing the details required for success in a rapidly changing market. Moreover, the demise of NCN revealed that the newspaper industry was less unified than many observers had believed. The industry staggered away from NCN as a disparate collection of companies with differing agendas and perspectives on the direction of the marketplace.

The developments recounted in this chapter—beginning with the debut of Mosaic through to the threats from new classified advertising competition—illustrate the fundamental shift away from the traditional media model. Operating within that traditional model, newspaper companies were central organizations that championed their role as agenda-setters and saw it as their mission to control when and how information would flow to readers. As was discussed in previous chapters, newspaper executives believed this industry's model, which established control from the point of content creation through to its distribution, could be translated into a successful online model. This premise underpinned the industry's partnerships with the proprietary online services.

However, newspaper companies found that operating on the Internet meant they could no longer control information flow; the distribution channel was no longer a one-to-many model. The Internet model was user-centric and embraced interactivity. Newspaper companies were forced to compete along with many other content providers as mere contributors to a vast, wide open network. While newspaper companies had no choice but to accept the new distribution realities—that was simply how the Internet worked—they struggled with relinquishing control over their content. The decision to embargo content—preventing it from appearing online until after it had appeared in print—demonstrates how seriously newspaper companies took this issue of content control at that time. The tension surrounding such content decisions defined the newspaper industry's relationship with the Internet in its early years.

During the late 1990s and early 2000s, the newspaper industry's Internet efforts were hampered by its efforts to address many cultural and business model issues. Many critics contend that by the time the industry understood that its model had to change, readers, advertisers, and investors no longer cared. Chapter 6 examines the newspaper industry and its relationship with the Internet during the first half of the 2000s. It pays particular attention to the newspaper industry's investment constituency and examines the reaction of newspaper companies to the Internet sector's investment bubble and subsequent crash. The chapter looks critically at how the newspaper industry viewed its own business model during this tumultuous period. In the early 2000s, investors demanded that public newspaper companies take action to validate their long-term viability. Chapter 6 first discusses the merger between AOL and Time Warner as a deal that reverberated throughout all sectors of traditional media and caused the newspaper industry to consider alliances with television stations as a pragmatic convergence strategy. Within this context, Chapter 6 discusses the ramifications of the newspaper industry's strained relationship with the investment community, and concludes with Knight-Ridder's decision to sell its newspaper holdings and close the company, which can be seen as the ultimate referendum on the newspaper industry's future.

6

Mergers, Convergence, and an Industry under Siege

AOL announced on January 10, 2000, that it planned to acquire Time Warner Inc. in a deal valued at $162 billion.[1] The monetary size of the deal attested to its scope and significance, but the transaction reverberated through all segments of the media industry as it became recognized as a seminal event: a new media company was about to buy a venerable old media company. The timing of the announcement—only days into the year 2000—served to amplify its cultural significance. The mainstream media, which had been rife with futuristic commentary surrounding the arrival of a new millennium, seized on AOL's announcement as evidence of fundamental change. One commentator within that atmosphere of rhetoric said AOL acquiring Time Warner signaled "a new era in both the culture industries and the economy more broadly," adding that the deal was "a rhetorical as well as financial watershed, the coronation" of the "new economy."[2]

Although the merger of AOL and Time Warner did not directly involve the newspaper industry, it represents an important milestone within the context of this book because it demonstrated how much clout new media companies had achieved by 2000. The AOL and Time Warner merger raised the stakes for traditional media companies trying to compete in the new marketplace. Many commentators described the merger as another development leading to the impending "death of the old media."[3] Newspaper industry executives were forced to react to the deal and to explain how their traditional media businesses could prosper in an era that was being defined by emerging new media brands.

This chapter examines how the newspaper industry responded to the market challenges confronting it in the first few years of the 21st century. It establishes historical context through a discussion of the economic climate of the period, which was defined by the inflation and subsequent collapse of the investment bubble in Internet-related companies. The chapter explores the intense media merger activity that took place during the period and presents the combination of Tribune Co. and Times Mirror Co. as an example of the newspaper industry's foray into the merger frenzy. The chapter also looks at how the newspaper industry, which had been confounded by the rise of Internet company stock values, assumed the dot.com crash represented weakened competition from new media. During much of the early 2000s, newspaper companies sought strategic solutions through partnerships with old television rivals as both forms of traditional media attempted to navigate a way forward. These partnerships led to many operational convergence initiatives. Media General's convergence project in Tampa, Florida, is presented as a detailed example of this trend. However, since these convergence projects usually failed to deliver tangible results, the chapter also explains how industry investors began to demand that newspaper companies articulate a long-term vision for their economic viability.

Throughout the 1980s and early 1990s, the newspaper industry believed that it would lead the media's digital transformation or greatly influence those who would. Yet the newspaper industry did not achieve its leadership goal, and by the early 2000s, was at risk of becoming marginalized as new companies transformed the Internet into a media platform. The example of Knight-Ridder Inc. illustrates how the fortunes of the newspaper industry had changed. Once at the forefront of the newspaper industry's online endeavors, Knight-Ridder faced intense pressure from investors who believed the company was underperforming. The investors urged the company to sell its newspaper holdings. When Knight-Ridder's management capitulated, the episode became symbolic of an entire industry's plight. Therefore, this chapter provides important insights into the critical period in media history when the U.S. newspaper industry collectively realized that its era of media dominance had ended.

The Internet Bubble and Collapse

In the early 2000s, newspaper company decisions regarding an Internet strategy were made against the backdrop of a capital market that underwent dramatic changes in a short time period. A vast escalation in the value of Internet-based businesses that began in the late 1990s came to an abrupt end in mid-2000. A period now recognized as the Internet bubble gave way to the dot.com investment collapse. Although not intended to recount all of the complex investment issues of this period, this brief description provides historical context.

Stocks in Internet-based companies soared in the late 1990s as investors speculated on the long-term potential of these businesses. As an investment class, Internet companies were not profitable, and most had unproven business models and uncertain sources of revenue. Nevertheless, investors bid up the prices of shares in Internet companies in hopes that their business models would someday align with their potential. The index measuring the value of the market where most Internet stocks were listed peaked at over 5,100 points in March 2000,[4] which represented a remarkable increase from less than 800 points in early 1995.[5] This surge in stock value gave Internet companies currency to make acquisitions that their balance sheets otherwise would not have allowed.

Gershon and Alhassan observed that the fascination with the AOL and Time Warner deal—beyond the huge dollar value—was due to AOL's role as the purchaser:

> What was particularly unique about the deal was that AOL with one fifth of the revenue and 15% of the workforce of Time Warner was planning to purchase the largest [media] company in the world. Such was the nature of Internet econom- ics that allowed Wall Street to assign a monetary value to AOL well in excess of its actual value.[6]

While Prodigy and CompuServe faded from prominence, AOL had exploited the Internet to its advantage in the late 1990s and built a subscriber base of more than 22 million by 2000.[7] The financial markets during this period of exuberance rewarded AOL by assigning an enormous value to the size of its audience. It was the market's valuation of AOL—not the company's underlying financial perfor- mance—that allowed it to acquire Time Warner.

Mergers Alter the Media Market

Although the AOL and Time Warner merger is recalled as symbolic of its time, it was part of a series of significant mergers that altered the media landscape. General Electric had acquired NBC, Disney had taken over ABC, and Viacom had pur- chased CBS. Even Time Warner was the result of a merger of two media giants, which had subsequently purchased Turner Broadcasting and its flagship property,

CNN.[8] An industry analyst said the merger activity stemmed from the belief that consolidation "should create media platforms with the leverage and scale to introduce [new] services widely and economically."[9] During this period, achieving economies of scale was viewed as a key business strategy within the media industry. Diversity of content was also seen as an important element that led media companies to acquire businesses involved in all types of media production.

Critics worried, however, that the level of consolidation would reduce the number of media owners to a point where the diversity of sources would limit consumer choice and threaten independent journalism by placing control of news organizations in the hands of only a few corporate owners. One analyst noted that some "now wonder how much more wheeling and dealing can go on before there are but one or two juggernauts controlling every image, syllable and sound of information and entertainment."[10] He concluded, however, that merger activity in the media industry had not run its course:

> Actually, the [media] industry has a long way to go yet before it reaches that point. There are more than 100 media companies worldwide, with more than $1 billion in revenues; and entertainment and media are still fragmented compared with other industries such as pharmaceuticals or aerospace.[11]

The issue was sensitive in the newspaper industry, where recent merger and acquisition activity had ended decades of family ownership and local market control. Newspapers were bought and sold at a record pace during the 1990s, and by the mid-2000s consolidation had changed the fundamental nature of the industry. An industry report found "the 10 largest newspaper groups already control half of the nation's daily circulation."[12] Nevertheless, industry analysts believed the newspaper industry needed more consolidation to achieve the economies of scale required to compete in the new media landscape. The circumstances confronting the newspaper industry prompted one analyst to observe:

> The rules have been changed by the globalization of business and, especially for media companies, the rise of the Internet. As uncomfortable as it might be for some, newspapers have no choice but to plunge in.[13]

The observation illustrates that, from the perspective of the investment community, the newspaper industry was not doing enough to remain viable. Newspaper executives were expected to take action.

A significant response came in the form of a large merger from within the industry. Tribune Co. announced it would acquire Times Mirror,[14] linking such stalwart newspapers as the *Los Angeles Times* and the *Chicago Tribune* under common ownership. Times Mirror owned no television stations (Tribune owned 22 television stations at the time) and had relatively little investment in Internet businesses aside from websites affiliated with its newspapers.[15] This meant that Tribune believed investing in additional newspapers was its best way to expand its geographic footprint.[16] Tribune's management said its acquisition of Times Mirror would give the merged company a sizeable presence in the nation's top three media markets and would allow the company to expand Internet operations there. Tribune's CEO stated:

> In the interactive market, we now have the spine of a national network. This should permit us to deal more effectively with whatever threats and opportunities arise. Tribune Interactive should now be a much more attractive candidate for partnership with any firm hoping to get a foothold on audience. And we should be far more attractive to the national advertisers that are the biggest players in the Internet.[17]

Critics had argued that the regional nature of most newspaper companies made it difficult for them to operate on a scale large enough to compete with media that offered advertisers national reach. Tribune's acquisition of Times Mirror was designed to address such market shortcomings.

However, the deal was criticized as a defensive move by old media businesses. One critical assessment said that "the combined [newspaper] company looks more like an industrial age holdout than a 21st-century media giant." The commentator called it "not too surprising that" Tribune's management "was spinning this deal as . . . a Web thing," but called it a "ploy to appease certain investors."[18] Whether or not it was a "ploy" is open to interpretation, but the fact that Tribune's management felt compelled to position the deal in terms of an Internet strategy reflected the underlying market changes. From an investment perspective, Internet companies had usurped power from traditional media businesses, which were forced to position their dealings in terms that this new market would accept.

Newspaper companies continued to post sizeable profits, with average operating margins of about 20 percent during this period. Industry executives believed that the market should reward such numbers, instead of rewarding Internet companies with enormous losses.[19] Although newspaper industry executives were confounded by what they perceived as the market's irrational behavior, there also was a strong interest by many to cash in on the Internet boom as it transpired. As was noted in the previous chapter, for example, several newspaper companies wanted NCN to issue public stock, and the disagreement over that issue was a factor in

NCN's demise. During this period, Cox Enterprises[20] and the New York Times Co. are examples of newspaper companies that formulated plans to spin out their online operations as separate public companies.[21] However, the stock market collapsed before any representative of the newspaper industry reaped any direct benefit from the investment bubble.

The Aftermath of the Bubble Collapse

By the end of 2000, the index measuring the value of NASDAQ, the market where most Internet stocks were listed, had lost more than half of its value and stood at less than 2,500.[22] Bontis and Mill concluded in a "post-mortem" analysis of the market during this period that in the absence of business fundamentals, investing in Internet stocks "was similar to gambling in a casino."[23] They found that investors acted more like speculators and assigned higher value to Internet metrics such as a website's number of unique visitors than to underlying business fundamentals such as the ability to generate revenue and control expenses.[24]

Despite their own missed opportunities to capitalize on the stock market's surge, newspaper executives collectively seemed relieved by the collapse of the Internet bubble. While no one embraced the recessionary economy that also affected this period, the collapse of Internet investment speculation was viewed by many newspaper executives as vindication for businesses that were profitable. These executives understood the market collapse did not mean the Internet would go away entirely, but they felt the sudden end to market speculation had validated their industry's cautious response. By not fully chasing the Internet based on investment speculation, the newspaper industry believed it had once again prevailed in saving its long-term franchise.

The following excerpt from the *Columbia Journalism Review* illustrates newspaper industry sentiment in 2001: "the rush to the Web was breathtaking. It began with bold proclamations that old media's days were numbered, and ended with old media—and its established brands—still standing when the smoke cleared from the dot-com burnout."[25] Not only were newspaper brands still standing, the industry was resolute in believing that it was poised to prosper in the next phase, as illustrated by the comment from the head of Knight-Ridder's Internet operations:

> We're seeing in the Internet industry what a lot of reasoned people have expected for a long time. After a phase one of innovation and experimentation, we're in a phase two of consolidation and seeing that some business models work and some don't. That is leading to a third phase, in which those who survive the second phase reap the considerable benefits of the Internet's growth.[26]

The AOL and Time Warner merger, hailed as a sign of the new economy in 2000, was by 2003 seen as evidence of the ruinous result of speculative investment. The combined company was in disarray. It reported a $99 billion loss, and a board of directors embarrassed by the outcome of the merger dropped AOL from the company's name. Gershon and Alhassan observed that "the AOL Time Warner merger may well be remembered as one of the worst mergers in US corporate history."[27] Newspaper industry executives were relieved to have avoided such a debacle and were intent on demonstrating to the investment community the wisdom of pursing profits rather than betting on speculative ventures. The newspaper industry believed the end of the Internet investment bubble represented an opportunity to win back favor in the investment community. By embracing profitability, newspaper executives wanted to show they knew how to create value where so many pure Internet companies had failed.

Newspapers Emphasize Online Profits

Against this backdrop, newspaper companies became headstrong in presenting their online editions to their shareholders as rational, profit-producing business units that enhanced the printed product. The publisher of the *Denver Post* expressed the sentiment of the industry, stating: "The future is print. Electronic communications strengthen print."[28] This reflected the prevailing attitude of newspaper executives during the period immediately following the collapse of the Internet investment bubble. Executives believed that newspaper companies had survived the arrival of the Internet and its associated irrational business practices. As such, they were positioned to use their companies' financial strength to exploit the Internet on their terms.

The cover of a summer 2002 issue of the industry's official trade journal, *Presstime*, underscored how important the issue of online profitability had become to U.S. newspapers. The illustration featured a man in a press hat carrying a bag of cash under the headline: "You Made Money!" The graphic depiction demonstrated an industry out to prove to its members and to others that newspapers could make money on the Internet. The article concluded that success was there for the taking: "No one can make the lame excuse that 'Nobody's making money online' anymore."[29] The amount of money being made, however, was a matter of interpretation. Some organizations claimed online profit margins of 40 percent or more. An industry analyst stated that "about a third of online newspapers broke even . . . a third were in the red and a third were profitable."[30] Nevertheless, the fact that two-thirds of the industry claimed to have figured out how not to lose money with their Internet operations was hailed within the industry as a significant accomplishment.

In looking back on this period, the New York Times Co. can be seen as having a significant influence on its industry's perspective regarding Internet profitability. As noted earlier, the company had attempted to capitalize on the skyrocketing valuations with a plan to make its Internet operations into a new company that would have gone public through an initial stock offering. When the market collapsed, those plans were shelved and a new strategy was adopted.

The New York Times Co. believed Internet operations had to become cash-flow positive in their own right to prove that the projects were worth pursuing. As the executive in charge of Internet operations stated: "we knew that just getting to break-even was not enough."[31] Achieving profitability, however, was not possible through revenue growth alone. The New York Times Co. cut expenses, which included eliminating 116 positions. In about a year, the company's digital operations went from posting a loss of $7 million to reporting a small operating profit of $200,000.[32] The performance was considered a newspaper industry milestone—Internet profitability had been achieved at one of its most stalwart companies. However, the industry's focus on Internet profitability led to increased scrutiny from analysts and investors who were interested in how those profits were calculated.

One analysis, for example, speculated that it was "doubtful that the *New York Times* could survive separately as a web edition," adding that "every profitable web news site of any significance depends on a non-web news organization, drawing on, but not paying for, its newsgathering resources."[33] The concern among industry analysts was that newspaper companies were not accurately reflecting the true costs of doing business on the Internet. The analysts were not as concerned with online profitability as the newspaper companies were, but they wanted public newspaper companies to be transparent in how they accounted for their online operations.

An executive with the New York Times Co. conceded that some newspaper "companies have a history of burying losses in new-media units—not deliberately, but just saying, 'We won't count this or we won't count that.'" But the executive said of his company: "we count everything," pointing out that the online operation paid 10 percent of its revenue, or at least $5 million annually, for the print content it used.[34] How newspapers addressed this issue varied from company to company. It was determined by an operating philosophy of whether or not Internet operations were seen as an extension of the printed product or as a separate business.

An executive of the *San Diego Union-Tribune* said "the fundamental idea is to carefully measure the incremental costs and revenues of the online operation," not to determine how the Internet would perform as a standalone business, "but whether it is a profitable new edition of the newspaper." The *San Diego Union-Tribune* attempted to assign most costs, but noted that some were not calculated as part of the online operations' expenses such as employee medical benefits or the value of promotion the online version received from the printed paper.[35] During this period, newspaper companies were comfortable in describing their online newspaper as an edition of the printed newspaper. It was a convenient metaphor for explaining organizational structure and the assignment of costs.

However, only a couple of years after the collapse of the Internet bubble, concerns over the minutiae of online profitability gave way to a much bigger issue. The newspaper industry continued to experience advertising erosion, and investors wanted to know when—or if—the online editions would produce enough revenue to offset this trend. Despite the investment collapse in Internet stocks, the popular financial press continued to describe the Internet as a significant threat to the newspaper industry's revenue. One article concluded that "over the horizon loom some of the toughest tests for the industry since the advent of television."[36] Newspaper companies had prospered during the television era, but the financial community believed the Internet represented a new form of competition—one that would directly threaten sources of advertising revenue that television had not. Given the mounting pressure to find alternative ways to operate, newspaper companies once again turned to a partnership model. This time, however, they were joining forces with their old television rivals under the guise of an emerging buzzword—convergence.

Convergence as a Strategy

Against the backdrop of media mergers in the early 2000s, Fallows stated that the activity "intensifies the impression that all media are part of one big octopus-like conglomerate."[37] The observation was part of the growing discussion in both industry and academic journals during this period in which media mergers and the notion of convergence became entangled.

Dowling, Lechner, and Thielmann have defined convergence as "a process change in industry structures that combines markets through technological and economic dimensions to meet merging consumer needs," adding that convergence "occurs either through competitive substitution or through the complementary merging of products or services or both at once."[38] Dowling and Thielmann also observed that such activity could "affect industry structures as well as firm-specific managerial creativity."[39] They wrote:

> . . . if convergence is the significant trend, the basic form can differ on a spectrum between competition and complementarity—leading either to a conglomerate market or to the emergence of a new market or market segment.[40]

Within this framework of convergence, newspaper executives attempted to answer two important questions about their industry's position in the market. First, could newspaper companies remain financially robust enough to be relevant in a market dominated by large conglomerates? Even executives of newspaper companies already part of large corporations believed their industry had to prove its relative worth. Second, could newspaper companies exploit partnerships so that such deals could achieve the scale required to be competitive in the emerging Internet market? Following the intense period of partnerships with proprietary services, most newspaper companies initially created an Internet presence on their own. But in the early 2000s, newspaper executives sought new partners they hoped would help them create richer online content and share the financial risk of broadening the product concept. The following section explores how the newspaper industry attempted to answer those questions by partnering with television stations in local markets in a form of operational convergence. Dominick had explained this as occurring "when owners of several media properties in one market combine their separate operations into a single effort."[41]

Newspapers and Television: Old Rivals Cooperate

By 2000, nearly all of the nation's 1,500 daily newspapers had a website.[42] The question of whether or not a newspaper would have a website was no longer pertinent; that question had been answered by an overwhelming view that an "online presence is essential to their competitive success."[43] The issue was now one of execution, and newspaper companies were seeking approaches that would create scale in the markets they served, but at a reasonable cost. In this effort, newspaper companies fixated on the possibilities of partnering with their one-time rivals— local market television stations.

Scott noted that a prevailing sentiment during this period was that news sites on the Internet were "a supplement and a complement to the dominant print and broadcast news media," adding that a "considerable debate" existed "over whether the internet will prove to be a new medium at all or, rather, more simply serve as a better tool for distribution."[44] Traditional media companies understood, however, that the investment community valued the Internet products more than the core television or newspaper business in many cases. Therefore, traditional media rivals—newspaper and television companies—set aside their competitive differences and pooled resources as they looked for ways to make their online efforts viable. Dominick explained the advantages they were seeking:

> It saves money because, rather than hiring a separate news staff for each medium, an operation can have the same reporters produce stories for the paper, Web site, and TV station. In addition, each medium can promote its partners. For instance, the TV newscast can encourage readers to visit the Web site or the print newspaper.[45]

Several researchers studied this convergence phenomenon and proffered explanations about why it occurred. Scott concluded that during this period the term "convergence" described a "new strategy in the economic management of information production and distribution."[46] Singer was more explicit in defining this practical manifestation of convergence. She wrote that "it refers to what happens inside a newsroom, specifically to cooperation among print, television and online journalists to tell a story to as many audience members as possible through a variety of delivery systems."[47] Singer also noted that this form of convergence relied on "some combination of technologies, products, staffs and geography among the previously distinct provinces of print, television and online media."[48] Lawson-Borders wrote that it reflects "the realm of possibilities when cooperation occurs between print and broadcast for the delivery of multimedia content through the use of computers and the Internet."[49] The common theme that emerged from these researchers is that the term "convergence" was no longer restricted by industry participants to a construct reserved for defining the technical consolidation of media platforms. As used by industry practitioners during this period, the term represented the practical notion of removing expense and overhead from news operations by combining the newsrooms of newspapers and television stations and the creation of content distribution alliances over the Internet.

In this application, convergence was applied to the cooperative partnerships that were formed among media companies that had been long-term competitors, specifically newspapers and local television stations in the newspapers' markets. To some these partnerships were unexpected, and viewed as hasty and imprudent reactions to Internet-based media. Wirtz, however, observed that it was a natural market development for convergence to take place among existing market participants in this manner. He surmised that such convergence activity was to be expected "since the multimedia market does not have to be built up from scratch, but is much more a combination of extant elements and applications."[50] He believed that a variety of players serving different aspects of a media market would join together because it made economic sense to do so. Wirtz concluded:

> . . . it remains evident that hardly any single market participant can capture the multimedia market alone. An integration strategy minimizes risk within the process of capturing the market through an efficient bundling of resources and a dispersion of the capital raising responsibility among several market participants.[51]

From this perspective, the participation of newspaper companies in convergence activity can be viewed not as a new strategy, but as a continuation of the risk management approach that had led to earlier partnerships with cable television companies and proprietary online services. Newspaper companies had always wanted to exploit online media, but had long shown reluctance to pursue projects independently. Television affiliations appeared to give newspaper companies a new way to move forward with online initiatives.

In 2001 there were about 50 convergence projects fitting this new definition underway in the U.S., and more were in the pipeline.[52] According to Gentry:

> This is all in its infancy and it's happening because newspapers are seeing subscriptions declining and TV stations are watching viewers decline and they figure that if they can cross promote each other and share resources, they can attract new audiences and save money. . . . This is much more than a trend. It's a movement.[53]

Indeed, in April 2003, duPlessis and Li reviewed the content of the online newspapers associated with the largest 100 circulated U.S. newspapers. The researchers wanted to gain a greater understanding of the effects of what they referred to as cross-media partnerships, which they defined as providing content to newspaper websites through alliances with media such as television stations. This study found that 86 of the 100 newspapers studied had some form of cross-media partnership,[54] underscoring how widespread the practice had become.

With so many projects underway or planned, it is not possible to describe them all except to say that they broadly fit within the understanding of operational convergence. Singer, for example, wrote that convergent "processes and outcomes vary widely" from market to market.[55] She observed:

> For some, convergence emphasizes information sharing. For others, it involves newspaper reporters taping a voiceover for a newscast, or television reporters phoning in breaking news details to update a website. In a few markets, journalists

gather information that they turn into an immediate online story, a package for the evening news and an article for the next day's paper. Convergence can mean working in separate buildings—or at adjacent desks.[56]

Singer believed that because of the diversity of such activity, the process would be fluid for a long period. She wrote that her "study indicates there are many different ways to converge, and models will evolve to suit unique organizations, markets and cultures."[57] Lawson-Borders shared this notion and wrote that "convergence is not static, but rather a continuum in which organizations must select the appropriate medium or combination thereof to reach their goals."[58]

To gain further insight into what this type of convergence represented, it is useful to consider several examples of such activity underway during this period. Convergence projects during this period often included properties that were under common ownership. Belo Corp. in Dallas, Texas, sought to foster partnerships among the properties it owned there, including the *Dallas Morning News*, WFAA-TV, and TXCN, a statewide cable news operation.[59] Tribune in Chicago teamed the news operations of the *Chicago Tribune* and its WGN-TV along with CLTV, a local cable news service.[60] Media General's project in Tampa, Florida, was the most often cited example of convergence occurring under common ownership and will be explored in more detail later in the chapter.

Such activity, however, was not limited to large metropolitan markets. The World Company in Lawrence, Kansas, for example, used its common ownership of the local newspaper, a television station, and the cable system to launch numerous cross-platform projects. The initiatives were so widespread that one critic described the small market as "the land that antitrust [law] forgot."[61] Projects also occurred outside of common ownership. For example, the *Daily Oklahoman* joined forces with Griffin Communications' KWTV to produce news stories for use in print and on television. Three months after beginning their traditional media cooperation, the two Oklahoma City media entities joined their Internet sites to create a new entity called NewsOK.com.[62]

Most analysis of the convergence projects from the early 2000s focused on content creation and dissemination. There was little mention of the business issues aside from cross promotion. In Topeka, Kansas, the *Capital-Journal* and WIBW radio, both owned by Morris Communications, partnered with KSNT-TV owned by Emmis Broadcasting, to not only share news resources, but to also sell "bundled advertising packages among the various media."[63] This partnership was one of the rare examples where advertising synergies were mentioned as part of the core strategy. As the general manager of NewsOK.com observed: "The editorial part was the easiest. In advertising and sales, no one really wants to share money."[64]

Therefore, most of these deals were designed around sharing expenses not revenue. The critical element driving this type of convergence was a fundamental assumption that leveraging resources among the various media entities involved would result in a competitive advantage.[65] It was implicitly understood that revenue would follow if such advantages were achieved.

In keeping with this line of thinking, some media companies believed their convergence projects represented another step in an inevitable evolution of media that would one day eliminate the distinctions separating print from broadcast. They pursued these projects out of a conviction that "breaking down barriers between media platforms, services and industries" was a transformative process that that would allow for once disparate industries to be "conceptualized under the umbrella of a new business sector, the information industries."[66] The following section presents a more thorough examination of the most prominent operational convergence project of this period.

Media General and Its Tampa Model

To understand the thinking behind convergence activity more clearly, it is helpful to look into the details of the way Media General went about the business of convergence in Tampa, Florida—one of the clearest examples of operational convergence in this period. Although as noted previously there were numerous convergence projects underway, the initiative undertaken in Tampa by Media General attracted the most attention from industry press as well as academic researchers. This was due to the size and scope of the effort relative to other projects and was reflected in the significant amount of resources Media General dedicated to the initiative. This example provides rich insights into how a traditional media company viewed its market during this period. Kolodzy had observed that if "convergence means cooperative relationships between television, online, and print media," then success would depend on management's ability "to play to the strengths of each medium, and to respect those strengths."[67] From the outset, executives at Media General expressed similar views, maintaining that convergence would work when traditional media took the best of their operations and combined those strengths with the efficiencies of Internet distribution.

Media General owned the *Tampa Tribune*, a 238,000-circulation newspaper; one of the market's leading television stations, WFLA-TV; and Tampa Bay Online, which was on the Internet at www.tbo.com. In 2000, Media General moved all three of its Tampa media properties into a new facility—a 120,000-square-foot building it called "The News Center." But as word spread in the industry about how Media General planned to integrate its media holdings, some began calling the new facility a "temple of convergence."[68] The following excerpt from an article published by the Poynter Institute describes the operations:

> It sits on the banks of the Hillsborough River in Tampa, a gleaming new $34 million building that has become the poster child for one of the most powerful but controversial trends sweeping the news industry: Convergence. . . . From the Internet to new breakthroughs in digital imaging to a public that demands better, fresher and more diverse news, the converging of different journalistic disciplines is dramatically changing the landscape of American journalism. Nowhere is this more ingrained than what's happening in Tampa. . . . TV reporters do their stand-ups and then write bylined newspaper stories. Newspaper reporters write their stories and then appear before TV cameras to do "talk-back" debriefings or their own stand-ups. And everybody—reporters, editors, photographers— "repurpose" their work for the website.[69]

Initially, management emphasized that although the news operations of the TV station, newspaper, and website shared a common physical location, they were continuing to operate as separate and distinct entities that "make news decisions independent of one another."[70] As the newspaper's editor stated: "we are careful to stress that there is no merger of the newsrooms, although the cooperation is unprecedented."[71] To many observers, however, the distinction between a "merger" of newsrooms and "unprecedented cooperation" was semantics. In practice, the centerpiece of the facility's new newsroom was an area known as the "superdesk," where newspaper editors, television producers, and website managers came together to coordinate the activities of all newsgathering personnel.

The superdesk was a tangible symbol of convergence. One trade journal stated that it "signals a change in the way journalists at all three media organizations will work."[72] Once the newsroom and its superdesk were operational, management conceded that many cultural issues had to be adjudicated. This prompted the newspaper's executive editor to state: "Convergence is a contact sport. It happens one staff collision at a time."[73] Singer's research of convergence, which included the Tampa market, concluded that the cultural issues of combining operations would be difficult to overcome.

The independent and competitive culture of newspaper newsrooms made it difficult for those journalists to accept convergence. Singer's study stated that print journalists especially expressed "being appalled when they learned they would be converging with their television counterparts."[74] Despite the concerns of many of his newsroom colleagues, Gil Thelen, the *Tribune*'s editor, believed the goal of convergence was worth the pain associated with forcing cultural changes into the organizations. In an article written for *The American Editor*, Thelen described for journalists why he supported operational convergence:

> Our purpose is to serve the changing needs of readers and viewers. They are ahead of us in using a combination of print, broadcast and the Internet during the day. Our rationale: Be there with news and information whenever and however our customers need and want us to be. For breaking news, we aim to "publish" on the first available platform, usually television but sometimes online. On enterprise, we want to extend the work of our journalists across platforms in a natural way.[75]

This comment was intended to help explain why a journalist would support convergence, but the editor also was aware that critics believed convergence was not an altruistic strategy, but one designed only for cutting costs in the face of increasing competition. Kolodzy summarized this perspective: "critics equate convergence with a loss of jobs, heavier workloads for journalists, and monolithic news and opinion. They see it as the manifestation of the dark side of media consolidation."[76] The Tribune's editor argued, however, that efficiencies associated with convergence would produce cost savings that would be invested in expanding, not shrinking, news resources. He stated: "If convergence leads to fewer journalists reporting, producing and editing weaker journalism, we deserve to lose customers and public trust."[77] Many media companies saw convergence as an organizational approach to confronting new market realities, but the journalistic ramifications of such organizational structures received much of the attention.

The changes were most troublesome for the newspaper reporters who found themselves in front of a camera for the first time. The comments of a senior reporter at the *Tampa Tribune* reflect these concerns: "The very nature of going on TV is intimidating for those of us hiding behind the anonymity of the byline . . . . I like to put a lot of thought into what I write. So thinking quickly [on the air] concerns me. . . . If I screw up, I can't backspace."[78] The experience of print reporters doing television reporting received significant attention in Tampa and led Media General to provide extensive cross-training through the University of South Florida.[79]

Lawson-Borders found that as editors and producers engaged in the converged process, they became more comfortable with using content across multiple platforms. A news director from WFLA in Tampa stated: "one of the basic truths about convergence is that not every story . . . that excites one platform is suitable for another. Sometimes a good newspaper story is just that [and] not suitable for TV."[80] Lawson-Borders concluded that "convergence for the sake of convergence is not advisable," adding that "the blending of media forms should be the strategy when the content and the delivery programs necessitate the arrangement."[81] The experience in Tampa, however, demonstrated that executive mandates were needed to force content sharing. Otherwise, the material was largely deployed exclusively in the medium that created it.

Even though it was the emergence of the Internet that gave newspapers and television a reason to try convergence in the first place, Media General's Tampa initiative underscored how difficult it was for traditional media to integrate the Internet into their operations. The News Center in Tampa was supposed to have been organized around what was described as the triumvirate of newspapers, television, and Internet. However, an early report on the converged structure indicated the Internet operations were not on equal footing. The report stated that "the television and newspaper voices dominate the convergence conversation, with the online operation looking for a place to fit in."[82] The online operation believed that the converged structure eventually would provide more resources, with an observer stating: "the TBO.com operation welcomes the opportunity to go beyond 'shovelware,'" and develop "more creative news presentations."[83] Thelen, who was considered to be an architect of the project, stated: "We want to place high value on experimental risk-taking, rather than on the tried and true journalism story."[84] However, the cultural challenges of moving beyond content repurposing were daunting and resulted in little innovative activity during the first year of the converged structure. Thelen conceded:

> We have only scratched the surface of multimedia possibilities. Effective translation from one platform to another is an achievement. But we have to learn how to create unique content that stands apart from any existing medium. That's the tantalizing prospect of multimedia work.[85]

This executive said the purpose of the Internet in a converged structure is to provide a central platform from which newspapers and television can both operate. He asserted that "online is the common carrier between print and broadcast," but said the purpose of the integration is to create something apart from what traditional newspapers and television already produce.[86] For Media General, developing this new content form was seen as the quintessential factor in determining if its convergence effort would succeed or fail, but the breakthrough it hoped for never came to fruition. Media General did not abandon its convergence approach, but the results were far less than what it envisioned when the strategy was embraced in 2000.[87] Media General backed away from aggressive attempts to merge content, and focused more attention on selling advertising across its three platforms. The company reported in 2005 that $8 million in annual revenue could be attributed to cross-selling opportunities among the newspaper, television, and online properties. For a market the size of Tampa, however, the number was small relative to overall advertising spending, representing "one-half of a percent of market share."[88]

Operational Convergence Fails to Deliver

Media General's experience in Tampa is important to understand because the experience soon turned out to be common across the industry. Operational convergence partnerships did not produce any significant results that industry executives could point to as evidence that traditional media had figured out a strategy for the emerging online market. Gordon suggested that most newspaper and television alliances were promotional in nature and not overly strategic.[89] As such, newspapers and television stations struggled to find ways to share traditional content, while at the same time, the online component of these arrangements was relegated to repurposing material intended for the traditional outlet. The notion that combining newspapers and television stations would result in new and interesting multimedia content packages was never fully realized.

In most cases, newspapers and television stations did not adequately address the cultural differences between organizations and among the various roles that would be required to cooperate. As Scott noted: "Journalists have been resistant, and coordination has proven labor intensive."[90] Singer observed that within the operations of traditional media companies, innovators—"individuals interested in doing something new largely because it is new"—were found mostly in the online operations.[91] But in the power struggles of converged television stations and newspaper companies, such individuals had little clout. Within the context of the Tampa model explored above, these individuals were referred to as internal "entrepreneurs," but there was "an ongoing concern [about] . . . how to integrate the entrepreneur into a traditional culture."[92] Management's inability to address the concerns of journalists, while also fostering an entrepreneurial environment, undermined the convergence strategy.

In the absence of major success, most of the newspaper and television partnerships forged in 2000 and 2001 were allowed to simply "fizzle" a few years later.[93] For example, researchers studying a convergence project in Oklahoma City in July 2002 found little cooperation between the partners and "the scant amount that was observed . . . seemed to be poorly coordinated."[94] In some cases, the convergence partnerships were reduced to promotional deals where, for example, a television station would provide its newspaper partner a weather forecast in exchange for logo placement.[95]

Some media companies such as Cox Enterprises formally announced plans to abandon operational convergence initiatives. Cox Enterprises had been a pioneer of the concept, launching Cox Interactive Media in 1996 to manage centrally all of the websites of the company's newspapers, radio stations, television stations, and cable systems. In markets where the company did not own multiple media properties, it formed alliances with outside companies. In early 2002, the company began to dismantle the central structure and return oversight of online operations to the traditional media business units. Cox Interactive Media attempted to operate on the Internet as a separate business, but was dependent on content and advertising relationships provided by the traditional media businesses. A senior manager of the project said "it wasn't a financial model that worked."[96] He elaborated: "We still think there's a tremendous opportunity for advertising," but added that "we need to have more participation from the local media companies to succeed."[97] Cox management believed that in separating its Internet operations into a separate unit, the company had failed to provide its traditional media businesses with the proper incentives to cooperate and build a base of online advertising revenue. In allowing each traditional media to control its own Internet presence, the hope was that ownership would provide the business incentives needed for success.

With operational convergence projects largely discredited as a way forward— and no clear replacement strategy on the horizon—media holding companies, including those that owned newspapers, began describing their approaches in several different ways. Some promoted a "portfolio" concept that referred to the core newspaper, its online edition, and other niche products such as a Spanish-language edition or commuter tabloid.[98] Other newspaper companies used "integration" to describe how their online editions and print editions were closely aligned under common management.[99]

For example, the *New York Times* in August 2005 announced a merger of its print and online newsrooms and also said that advertising sales for the print and online editions would be managed jointly under a new "chief advertising officer." The executive appointed to that role said the structure made it clear that "there's no us and them. There's no print and digital. We're all the *New York Times* Company."[100] The close integration of digital and print operations was intended to show advertisers and investors that newspaper organizations recognized the importance of the Internet, but in a way that inextricably linked online to the core printed newspaper. However, they largely were seen as semantic exercises rather than expressions of new strategic approaches. Operational convergence had at least represented the notion of reaching beyond a newspaper's own resources to create something new for the Internet. These new approaches, however, represented a return to the newspaper industry's reliance on its core printed product, with little sense of the novel opportunities that interactive media online might provide.

Lind argued that the failure of convergence projects to produce business results was largely because they were born out of the rhetoric surrounding the Internet rather than from any conviction that they were needed or wanted. By searching newspaper databases for this period, Lind found that usage of the term "convergence" spiked in 1994, waned for a period, and then rebounded for "a second broader peak during the Internet bubble 1999–2001." Lind said the usage patterns for convergence followed a model called the "Hype Cycle" developed by Gartner Inc. to explain in part what happened during the Internet boom.[101] Lind surmised that the use of convergence "gives few guidelines for concrete strategic action."[102] He concluded that convergence was "a general and rather vague term imbued with a lot of hype" and "seems to have been used as rhetoric and a residual argument, in lack of more concrete strategic arguments."[103] With this perspective, it is possible to understand how some newspaper executives and their television counterparts thought of convergence in nebulous terms that resulted in partnerships where goals and success measurements were ill defined.

Other research found little business justification for cross-media partnerships. Jung studied product diversification at 26 large media companies from 1996 through 2002 and determined that benefits assigned to media mergers, diversification, or integration were a myth. "The mantra of synergy does not work in the media industry at this point," he wrote, adding that data showed the opposite in that "financial efficiency decreased as the firms expanded their businesses into unrelated media sectors."[104] Jung asserted that the operations of newspaper and broadcast companies were too different for any pooling of resources to be effective. He concluded:

> . . . relatively few opportunities to make better use of collective resources will arise directly from related diversification of these particular sectors of media. If economies of scope are non-existent and financial profits are generally difficult to achieve, few economic benefits can be directly attributable to cross-ownership of television and newspapers. . . .[105]

Jung's work shows there was little business rationale for the newspaper industry's convergence projects with television stations. Therefore, when these convergence projects are considered within the historical timeline, they are recognized as a part of the hyperbole associated with the Internet bubble. *Advertising Age* observed:

"when the dot-com bust put the kibosh on so many media dreams—convergence became discredited."[106] Newspaper companies and other traditional media used convergence as a response to an investment community inflamed by the Internet investment mania. However, it was a response driven by wishful thinking that such partnerships could produce a breakthrough. There was little, if any, strategic planning or business planning undertaken to support that supposition.

In the wake of the Internet investment bubble and collapse, newspaper executives believed they had the opportunity to use their profitable businesses to regain standing with the investment community. Strong profits from print operations were supposed to bolster the industry's value, while operational convergence initiatives demonstrated a path to a sustainable future. However, an economic recession in the early 2000s exposed weaknesses in the industry's core product, and the convergence shortcomings revealed a lack of long-term strategic planning. The following section explores how the newspaper industry's emphasis on profitability affected its strategic decisions during this period.

Business Model Questioned

Newspaper executives considered robust profits a key factor for evaluating performance in the tumultuous media market of the early 2000s. However, the emphasis on profitability created a conundrum for the newspaper industry. In delivering large profits, newspaper companies conditioned investors to expect them. But at the same time, the investment community expressed concerns that newspaper executives were focused on short-term thinking that kept the industry from doing more to compete against new and successful online companies that were siphoning away revenue. Some of the profitability attributed to online editions, for example, had been achieved through cutting staff and pooling resources with competitors rather than investing in new technologies and business models to increase revenue.

Some blamed the industry's decisions to cut resources on its reaction to the end of the Internet investment surge. As an industry commentator wrote: "the burst of the dot-com bubble . . . made many think they had overestimated the impact of the Internet. But in retrospect, the news media might have completely underestimated the influence of this new medium."[107] This line of thinking gained currency in 2004 and 2005 as the newspaper industry continued to post revenue declines in its print business, while revenue from Internet editions fell far short of making up the difference. Newspaper companies were attracting a sizeable number of users to their websites, but they demonstrated no coherent strategy for generating revenue based on this audience.

Critics of the newspaper industry's Internet efforts have often asserted that websites associated with newspapers were unappealing to consumers because they lacked the features and functions associated with pure online companies. By the mid-2000s, however, newspaper companies had mitigated many of those shortcomings. An analyst with a leading audience measurement firm noted that "most, if not all, of the top newspaper websites offer interactivity such as blogs, podcasts and streaming video/audio." The analyst said that such features, when added to the news content, "make newspaper websites an increasingly appealing choice."[108]

Consumers responded to the improvements newspapers made to their websites. Nielsen/NetRatings reported in October 2005 that newspaper websites experienced double-digit audience growth from the previous year and reached 39.3 million readers, which represented "26 percent of the total active American Internet population." Websites associated with two national newspapers were attracting large audiences. For example, nytimes.com had more than 11 million monthly users, while usatoday.com tallied more than 10 million users.[109] A media industry investment specialist also observed that newspaper companies were attracting enough online audience to be "the No. 1 or No. 2 local Web site" in a newspaper's market.[110]

At this point in the newspaper industry's relationship with online media, attracting audience on the web was much less of a problem than finding advertisers willing to buy that audience. As the newspaper industry's overall business model came under increasingly intense investor scrutiny during this period, newspaper publishers were expected to reduce their dependency on the core printed newspaper and find ways to sell the online audience to advertisers. As the investment specialist summarized: "that's part of the challenge, how do you capitalize on the assets you have?"[111]

A researcher who studied the newspaper industry's reaction to the Internet as a disruptive technology stated that newspaper companies had not done enough to differentiate their products for advertisers. Gilbert said that "most companies aggressively 'crammed' the new business into the old business model and sales processes." In terms of revenue generation, he found that "most newspapers tried to force their online sites to make money by selling the same types of advertising to their traditional print advertisers" without recognizing that "online advertisers were different and the type of advertising they sought was much more focused around the interactive and direct targeting attributes of the new media."[112] The result, as Gilbert concluded, was that "newspapers had spent a ton of money, with little to show for it."[113] This perspective illustrates another instance when newspaper companies did not understand interactivity in the way the market demanded interactivity. Although newspaper websites had improved their offering of interactive content after years of neglect, most newspaper websites did not extend that functionality to advertising where the market had moved.

Nevertheless, investors wanted newspaper companies to respond to their demands to fix the situation. They wanted to see tangible evidence that newspaper companies could attract online revenue in amounts large enough to offset the negative advertising trend in the printed newspaper. Newspaper publishers, therefore, were forced to accept online advertising revenue rather than overall profitability as the benchmark that mattered. By this point, investors in the newspaper industry had moved beyond evaluating companies on short-term profits. They were far more interested in the newspaper industry's long-term prospects.

In 2005, the newspaper industry reported that Internet advertising generated by newspapers' online editions collectively "topped $2 billion for the first time."[114] Although this represented 31.5 percent growth from the previous year, it still meant that online advertising accounted for less than 4.5 percent of the industry's overall advertising revenue.[115] When Google, as a single entity, reported $3.2 billion in online revenue[116]—$1 billion more than the entire newspaper industry produced online—publishers began to realize how dramatic the shifts in the media market actually were. That realization—when combined with the long-term erosion of advertising and circulation in printed newspapers—created an atmosphere of failure.

The newspaper industry's standing with investors suffered further as the veracity of its reported online revenue met with skepticism. Given that as much as 70 percent of newspaper industry online revenue was attributed to classified advertising,[117] analysts believed that most of the reported online revenue represented upsells from the printed product. The revenue was viewed as so closely tied to the printed newspaper, the industry was given little credit for creating any traction in the online advertising market. Instead, the newspaper industry was seen as losing ground to several Internet companies. Google's introduction of simple text advertisements appearing alongside its search results was seen as an innovation that outflanked newspaper companies. Google had emerged as the best of the search engine category by deploying complex algorithms to answer specific user inquiries. Using targeted advertising to create value out of the seemingly endless number of search results represented a new form of advertising that was not in keeping with a media model.

A Google executive explained that traditional media companies approached online advertising with their offline model by "packaging content or advertising inventory."[118] This model, however, implies scarcity such as the limited number of pages in a newspaper or the finite amount of time in a television program. On the Internet, potential advertising inventory is endless, expanding as the audience increases. Therefore, Google did not approach solving the online advertising riddle by thinking in traditional terms. The executive stated: "We look at ads as commercial information" and present them as part of "our core mission of organizing the world's information. When people in the media world hear this, they say, 'What are these guys talking about?'"[119] The newspaper industry's inability to understand Google's approach as well as the strategy of other newcomers made it difficult to mount a competitive response. The statements of one leading industry executive suggested that the industry's predicament was rooted in complacency.

Speaking in 2000 about the emerging online competition in classified advertising, the chief executive officer of Times Mirror was very candid: "You'd think we'd know everything about [classified advertising] because it's so important, but we actually know remarkably little, because, I think, we as an industry have taken it for granted."[120] As this perspective became more understood in the financial community, investors became increasingly alarmed that the newspaper industry was unprepared to stave off the competitive threat posed not only by Google and Yahoo, but by Monster.com and other start-up companies that had taken direct aim at the newspaper industry's vital revenue stream from classified advertising.

A 2001 academic study of online classifieds associated with 75 newspapers found that publishers "have not developed their online advertising classified sites to take advantage of existing technologies for delivering effective, interactive classified ads."[121] This lack of a technically competitive platform contributed to the newspaper industry's failure to keep pace with new market entrants. An industry consulting study found that newspaper companies lost 5 percent of their classified market share to pure Internet companies such as Monster.com, Realtor.com, and Craigslist from 2001 to 2004. The erosion represented nearly $2 billion in revenue.[122]

Gilbert compared the income statements of new start-up online businesses with the income statements of online newspaper businesses, revealing "that anywhere from 35 percent to 45 percent of the categories of revenue are missing from the online newspaper." He cited Monster.com as an example of a start-up company finding new streams of revenue from employers placing recruitment advertising. He asserted that Monster.com was a direct threat to the newspaper industry's core classified advertising market, noting that it had posted more than $100 million in net income during the period of his research. "There is real money being earned all around these [newspaper] guys, but they continue to insist that the market doesn't exist."[123] Gilbert concluded that newspaper companies were missing the larger perspective by operating their online businesses as close extensions of the printed newspaper. Newspaper companies, he argued, need "to recognize that this is a new business. If you start and stay integrated, the online income statement will always look like the newspaper's."[124] Internet-based competition soon understood there was little to fear from a newspaper industry that showed no signs of mounting a serious strategic response.

The chief executive officer of Monster.com, for example, observed of the newspaper business: "You have a lot of jobs, a lot of ink, rolls of paper, unions, printing presses, trucks, offices, all of them being supported by the way the newspaper has run for 100 years."[125] The implication was that newspaper companies were too entrenched in established processes to separate their online and print products. Other observers echoed similar sentiments with one stating: "The newspaper and online in their minds are Siamese twins."[126] Once again, the lack of tangible financial results with online media opened up the newspaper industry for criticism about its approach to the market. Unlike earlier periods, however, the industry's long-standing record of profitability in its core printed business failed to serve as a counter argument, and investors began to abandon newspaper stocks as they questioned the industry's long-term viability.

A Wharton Business School professor put it bluntly, stating the newspaper "industry has matured to the point to where it has been a little lazy."[127] The *Wall Street Journal* stated during this period that newspaper companies "face an image problem," in that "they seem slow and stodgy when compared with some of their media rivals—namely, cable [television] and the Web."[128] Newspaper companies were described as lacking the necessary economies of scale amid "a renewed push toward consolidation" and the industry was chided for becoming less relevant to advertisers.[129] An advertising consultant summarized the circumstances facing newspapers: "The challenge for newspapers is . . . to figure out what they can provide that isn't being provided by the Internet and CNN."[130] Some newspaper executives attempted to find strategic answers by acquiring Internet businesses, a process seen as a logical move for companies operating in the mature newspaper industry with the financial wherewithal to close such deals.

For example, the New York Times Co. acquired About.com for $410 million; Dow Jones & Co., publisher of the *Wall Street Journal*, bought MarketWatch for about $500 million; the Washington Post Co. purchased

Slate, an online magazine, from Microsoft; while Gannett Inc., the Tribune Co., and Knight-Ridder Inc. each acquired a 25 percent interest in Topix.net, an online news aggregation service.[131]

Such acquisitions are representative of the "inter-media struggle for survival in the Internet age" as presented in a framework developed by Lehman-Wilzig and Cohen-Avigdor. The researchers found that acquisitions are part of "the natural life cycle of new media evolution" and are adaptive tactics for traditional media seeking to prolong their demise.[132] They wrote that acquisitions can provide traditional media with new sources of revenue to "subsidize" their transition into new media, but they noted: "Of course, this does not guarantee older medium survival."[133] The examples cited are noteworthy given the strategic importance they represented for the acquiring newspaper companies. However, most newspaper companies did not embrace an acquisition strategy as part of a long-term Internet plan for a variety of reasons.

By this time, the problems with the AOL and Time Warner merger cast doubts on the viability of any merger between traditional media and new media companies. Also, the market value of newspaper companies was declining as investors turned away from newspaper industry stocks, which made it more expensive for newspaper companies to use stock to fund acquisitions. Furthermore, most Internet companies saw no reason to be acquired. Aided by Google's successful initial public stock offering in 2004, Internet stocks were rebounding in value after the meltdown earlier in the decade. The market perception had shifted again: newspaper companies needed the Internet, but the Internet did not need them.

The Perspective Changes

Newspaper executives had assumed that the industry's financial strength and market position would provide them with the means to compete against anyone. However, as long-term trends in the core printed product turned increasingly negative and competitive threats from Internet competition intensified, investors demanded change, but they also grew more doubtful of the newspaper industry's ability to deliver it.

Newspaper companies remained profitable in 2005, but margins were "still well off the peak of 1999 and 2000."[134] An industry participant, in discussing the changing economic structure, observed that newspaper executives "get it intellectually. But they struggle with the emotional issues and the financial dynamics. It's really hard to cannibalize yourself and trade high-margin revenues for low-margin revenues one second before you have to."[135] But the profit erosion of the early 2000s was enough to convince industry leaders that the market had shifted in dramatic ways. The president of Cox Newspapers, which published the *Atlanta Journal-Constitution* and 16 other daily newspapers, stated:

> I think we were slow to catch on. I think it's perfectly natural to protect what you have; to think what you have is the only thing that people want. You look back to the year 2000, and I don't know that newspapers ever had a better year financially. Those are the times that can lull you into a sense of complacency. What we've discovered . . . is that that world doesn't exist anymore.[136]

Understanding this new business reality, the newspaper industry's trade group joined the chorus for reform. It issued a report titled "Why the Current Business Model Needs to Change," in which the industry was challenged to find new ways of operating in order to absorb a projected 20 percent decline in advertising revenue over the next five to ten years.[137] The industry's own projection for such advertising erosion was another reminder of the challenges confronting newspaper executives.

Within this context, the story of Knight-Ridder Inc. is an example that more fully explains the severity of the financial circumstances surrounding newspaper companies during this period. In November 2005, Knight-Ridder executives surprised their industry colleagues with an announcement that they were putting the company up for sale. As the *American Journalism Review* described it, "Wall Street's dissatisfaction with newspapers boiled over" as frustrated money managers and investors "forced" Knight-Ridder's management to sell the business.[138] The following section explores the circumstances surrounding this significant episode and explains the company's willingness to give in to the dissident shareholders as symptomatic of the overall industry's predicament.

Knight-Ridder's Demise

Tension between Knight-Ridder's management and several investor groups had been building since April 2005. The investors were unhappy that Knight-Ridder's stock price did not, in their view, reflect the true value of the business, and they openly questioned the company's ability to execute a strategy that would change its value proposition.[139] A newspaper analyst reported that Knight-Ridder's profits had failed to keep pace with an overall industry going through deep cuts. In 2004, Knight-Ridder's operating margin was 19.4 percent, which was close to the industry average of 20.5 percent. In 2005, the industry average declined slightly to 19.2 percent, but Knight-Ridder's operating profit fell to 16.4 percent.[140]

With such performance as a backdrop, dissident shareholders believed the company's parts were worth more than the entire company. In early November, the chief executive officer of an investment group that owned about 19 percent of Knight-Ridder's stock issued a letter that called for the company's sale because of the "significant and persistent disparity" in the price of the company's stock and the investors' perceived value of the company.[141]

Knight-Ridder's managers had taken several actions to appease their unhappy investors including significant layoffs, but those efforts fell short of making any impact on the stock price. The company's shares fell about 14 percent between a July board meeting and the time the dissident shareholders issued their letter in November.[142] Less than two weeks after receiving the letter, Knight-Ridder's management conceded. The company announced it would "explore strategic alternatives . . . including a possible sale."[143] Early the following year, Knight-Ridder was sold to McClatchy Co. for about $4.5 billion, and McClatchy in turn sold a dozen of the newspapers it acquired to make the deal financially viable.[144]

Newspaper industry executives understood the investment climate confronting their industry, and they were aware of the pressure being exerted on Knight-Ridder management by the dissident shareholders. Nevertheless, the demise of the venerable company stunned many of them and left an indelible mark on the industry. A former Knight-Ridder publisher summarized the questions everyone in the industry was asking in the aftermath of Knight-Ridder's closure:

> Could anyone imagine 10 years ago saying that in 10 years, Knight-Ridder would not exist? It was one of the strongest newspaper companies in America. How could you have a hand like that and play it in such a way that you would end up losing everything?[145]

The irony of Knight-Ridder's demise is that it was profitable and at the time of its closure was posting margins "higher than that of many Fortune 500 companies, including ExxonMobil."[146] As such, Knight-Ridder's management received some criticism for not fighting to keep the company intact.

In a letter to shareholders, chief executive officer Anthony Ridder explained: "I wish the solutions, as some have suggested, were only as simple as strong operating results or even just producing more great journalism."[147] He later added that Knight-Ridder's problems were linked to broader concerns about "what's happened to the newspaper industry over the last couple of years," especially its lack of revenue growth.[148] By the time Knight-Ridder was facing its group of activist shareholders, newspaper stocks were no longer priced based on short-term business fundamentals such as quarterly profits. The financial markets were ignoring the newspaper industry's current profits and bidding down shares based on the prevailing sentiment that newspaper companies could not sustain those profits for the long term.

When McClatchy emerged as the buyer for most of the Knight-Ridder newspapers, its chief executive officer maintained that investors were wrong about the newspaper business. He stated: "Pessimism about our industry is indeed widespread . . ., but we think it's misinformed. Newspapers remain profitable businesses with strong audiences."[149] Industry analysts observed, however, that McClatchy "was the only newspaper company to submit a formal bid" to purchase any of the Knight-Ridder newspapers.[150] One expert in media deals commented that "15 years ago, we would have seen 12 or 14 players bidding on these assets; the auction process would have been a frenzy."[151] The industry's critics asserted that the lack of interest in Knight-Ridder's newspapers shown by other newspaper companies was the greatest indictment against the future of the newspaper industry.

In retrospect, the other newspaper companies were proven correct in avoiding further investments in the industry. Subsequent to the deal, McClatchy's management received more criticism for the purchase than Knight-Ridder management had received when the sale was announced. For example, one commentator called McClatchy's purchase a "major error" and described it as "doubling down on the decaying industry."[152] Investors bid down McClatchy's stock after the deal, and its shares continued to decline as the overall newspaper industry sank as an investment sector.

Industry Ramifications

The effects of the Knight-Ridder closure on the attitudes of newspaper executives were significant. One official described it as a "tragedy" and said it was the industry's "real wake-up call" that prompted many executives to understand that "the wolf is closer to our heels than we thought."[153] A commentator wrote that "soul-searching in the industry has intensified,"[154] and another observed "the message is that newspaper organizations are going to have to change pretty drastically to hold onto their franchise."[155] Much of the commentary associated with assessing Knight-Ridder's closure reflected this internally focused rhetoric. But there were assertions that newspaper companies were not entirely at fault for their predicament.

This perspective emphasized Knight-Ridder's profitability and blamed its demise on greedy investors. As one commentator stated: "Wall Street only knows one mantra: 'More please, more.'" This observer said the case of Knight-Ridder illustrated that "we damn well have got severe problems with investors who in my opinion are completely unreasonable," adding that "I do not see that pressure lessening."[156] The rhetoric supporting this perspective was prevalent among editors and journalists. They often blamed investors for ignoring the public trust role of newspapers and were angry at newspaper management for attempting to appease those investors through budget cuts and layoffs.

A former Knight-Ridder editor described this philosophy of newspaper management as "the notion that you can continue whittling and paring and reducing and degrading the quality of your product and not pay any price."[157] This line of thinking continued to place the printed newspaper at the center of the industry's economic issues. Those espousing this view had not grasped that the financial markets were no longer valuing newspaper companies in the present; they were devaluing them based on expectations of unfulfilled potential in the Internet era.

That investors even noticed the short-term financial results was the fault of an industry that provided nothing else upon which it could be valued. Picard asserted that the newspaper industry's operating methods created a climate where "investors pressure them for short-term returns more than they do other types of companies" because those other businesses "are able to articulate a vision of a sustainable future."[158] Picard explained that, in the absence of a believable longer-term strategy, investors respond as investment rationale dictates:

> What these investors are looking for is a good return on their money; to get that they are willing to trade short-term profit for long-term growth and stability. But most publicly traded newspaper companies offer no credible plans (or vision) for anything beyond the delivery of higher-than-average quarterly profits. With this mentality in place, investors pressure boards and management for high returns so that they can recoup their investment in a shorter period of time.[159]

Once established between an industry and the financial markets, an investment pattern based on short-term results is difficult to bring to an end. Management becomes fixated on expense controls and ends up eliminating resources that could have been productive over a longer-term scenario.[160]

As this investment pattern played out in the newspaper industry, Rosen observed: "They won't stop the gravy train even though the engine is broken. How does such a thing eventually stop? It crashes."[161] Rosen's view of this "profitable demise"[162] was corroborated in Picard's economic analysis, which presented the construct as a downward spiral. Picard wrote that the newspaper industry's history of chasing short-term profits

> . . . abets uninterested investors by draining resources from newspapers they believe have a limited (or no) future and leaves newspaper enterprises without sufficient resources to renew themselves. The prospect of demise, coupled with the lack of strategic vision, becomes a self-fulfilling prophecy.[163]

To some observers, Knight-Ridder's closure represented a tangible example of this investment pattern reaching its ultimate conclusion. However, that is not an entirely accurate representation given the company's financial position at the time it was sold.

Despite several years of cost-cutting, Knight-Ridder was profitable and was positioned to remain so for many quarters. Therefore, this newspaper company did not "crash" as in Rosen's vernacular. Rather, the case of Knight-Ridder is an example of capitulation. Knight-Ridder's management could have continued its course until the company was indeed bankrupt. Management also could have attempted to break free of the investment cycle fuelled by cost-cutting demands through crafting and enacting a strategy for long-term viability.

An analyst with the investment banking firm Merrill Lynch said she had advised Knight-Ridder management to follow another course instead of selling the business: "I said, if you have the conviction that the news business and . . . online are a win, put up a 'work in progress' sign and say that margins are going to go down" until a new strategy is in place.[164] The company's dissident investors would have fought such an attempt to wrestle back strategic control of the business, but the decision not to engage in such a fight positioned Knight-Ridder as weak. That perception was transferred to the entire newspaper industry.

Knight-Ridder's chief executive officer responded to characterizations of his decision to sell the business as surrender by stating that "part of me would have loved to have had this fight," but said he concluded that the ensuing "turmoil" would not have saved the company.[165] Nevertheless, critics contended that Knight-Ridder gave up too easily and in the process harmed the newspaper industry by contributing to what Picard described as "a widespread sense that investors, as well as some newspaper owners, are giving up on the industry."[166] Therefore, Knight-Ridder's demise came to symbolize the state of the newspaper industry in the mid-2000s. Despite the positive outlook espoused by its acquirer, Knight-Ridder's inability—or unwillingness—to develop a strategy for long-term viability cemented the perception that the newspaper industry had squandered its opportunity to exploit the Internet and had passed into an era of long-term decline.

Discussion

The AOL acquisition of Time Warner at the outset of the 2000s appeared to herald a new era in media economics. It both reflected and also exacerbated corporate rhetoric around the rise of new media and the demise of old media. In the wake of the deal, the concept of convergence gained renewed currency, and traditional media companies began to see convergence as a strategy that might provide some means by which they could retain or improve profitability. Unsure of their position in the emerging new media market, newspaper companies sought alliances with their one-time nemesis—local market television stations. In some cases, these were partnerships forged among companies with common ownership. In other cases, they were deals between media outlets that had been intense rivals. The hope was that in combining the resources of print with the immediacy of video, these alliances would be able to craft a new media form for the Internet era.

When the Internet investment bubble collapsed, many executives in the newspaper industry felt vindicated when their companies were still standing, but many of the Internet companies they thought represented competition were no longer in business. Emboldened by the return of rational markets, newspaper companies sought to make their online operations profitable as quickly as possible rather than invest in long-term research and development. However, newspaper executives had misread the turn of events. The end of Internet investment speculation did not represent the end of Internet innovation.

As a practical construct, however, operational convergence as represented by newspaper companies and television station alliances did not meet the strategic demands of the Internet. In fact, most of those initiatives failed to produce any tangible benefits. Newspaper companies scaled back their convergence aspirations, and most of the alliances were relegated to the same history that included the failed videotext projects of the 1980s and the unfulfilled partnerships with proprietary online services in the 1990s.

Newspaper companies were alone again to face resurgent portals and an upstart named Google. By 2005, most newspaper websites were recognized for attracting a sizeable audience, but newspaper companies were criticized for their inability to generate revenue based on this audience. Investors wanted newspaper companies to explain how they planned to compete; they wanted executives accustomed to producing a new product every day to step back long enough to see past immediate shortcomings and articulate a coherent plan that would demonstrate the industry held long-term viability.

With the ability to earn a respectable long-term return on an investment in the industry in doubt, the financial community hammered newspaper companies for short-term profits. The result was a downward spiral fed by budget cuts and layoffs that drained resources and made the prospect of long-term viability even more unattainable.

Knight-Ridder emerged as a microcosm, exhibiting all of the turmoil of the larger industry. As dissident investors argued over the value of the company and its strategic direction, senior management capitulated. Rather than wage a contentious fight, senior management sold the company's newspapers and closed the business. The events surrounding Knight-Ridder epitomized an industry in decline, one weakened to the point that one of its most stalwart companies chose going out of business as its best strategic option. Gilbert's observation presented in the chapter underscored how newspaper companies ignored calls to treat online as a new medium. Industry executives were stubborn and arrogant in failing to adapt although—as this book has illustrated—they had ample time to do so.

Newspaper companies had assumed, correctly, for some time that a large portion of the advertising revenue from printed newspapers would one day need to be replaced with online revenue. The strategic error was in misjudging how soon that would occur. Throughout its history with digital distribution, newspaper companies never elevated their online endeavors to the level of strategic importance that would prepare them to take over the role as leading revenue generator when the tipping point arrived. The newspaper industry's ability to produce strong profits for so long established a sense of complacency that proved impossible to dislodge.

7

Connecting the Lessons of History

In August 2010, Gannett Co., a major publisher of U.S. daily newspapers, reorganized the operations of its flagship product, *USA Today*, as part of a plan to deemphasize the newspaper's print edition and increase resources allocated to electronic distribution such as Apple's iPad, other tablet devices, smartphones, and the web.[1] To some observers, the plan was a long-awaited admission from a prominent publisher that if the industry were to survive, the printed newspaper could no longer be the primary product focus. To others, however, the reorganization was merely another round of expenses cuts—130 jobs were eliminated[2]—by a company operating a dying business couched in rhetoric to appease skeptical investors. Newspaper executives had said many times before they knew what needed to be done to succeed in the digital era, but then changed little about how their companies operated.

This time, however, the newspaper industry was confronting another technology shift. Similar to the mid-1990s when the Internet emerged as a media platform, emerging tablet computers and smartphones were providing newspaper companies with an opportunity to re-establish their brands in the digital marketplace. Applications for distributing content to smartphones and tablets were seen as tools to attract paying consumers in ways that had eluded newspapers on the World Wide Web.

Chapter 5 explored the mid-1990s shift away from proprietary online services to the World Wide Web, as illustrated by an October 1994 article in *Wired* that asserted the advent of the browser made services such as Prodigy and AOL obsolete. Nearly 16 years later, the same magazine proclaimed another major technology shift was underway. In its September 2010 cover story, *Wired* declared "The Web Is Dead"[3] in a marketplace increasingly reliant on "platforms that use the Internet for transport but not the browser for display."[4] Forecasters project that the market for mobile applications will surge from nearly $7 billion in 2011 to as much as $35 billion by 2014.[5]

However, the importance of the mobile marketplace to newspaper companies transcends applications. The future of the industry depends on the ability of newspaper companies to capture advertising dollars in the mobile market as advertisers shift their budgets from print media products to mobile media products. Data compiled by eMarketer.com explained the circumstances confronting newspaper publishers. The 2011 data compared the time consumers spend with a particular medium and each medium's share of the advertising market. This analysis showed that television is nearly balanced with the medium commanding 42.5 percent of the time consumers spend with media and 42.2 percent of total advertising. The Internet was approaching balance with 25.9 percent of the time consumers spend and 21.9 percent of total advertising. Mobile media attracted 10.1 percent of the consumers' time, but captured only 0.9 percent of the advertising market. Newspapers also had an imbalance in the time and advertising comparison, but in the opposite direction. Newspapers attracted just 4 percent of the consumers' time, but received 15 percent of the advertising spending.[6] The analytics company suggested that print newspapers will be unable to sustain such an imbalance as advertisers ultimately follow the audience to mobile media:

> Shifts in ad spending remain behind the shifting consumption patterns of the US population. While TV is clearly getting its fair share of dollars, the amount of ad spending going toward digital does not yet reflect the amount of time consumers have invested in these areas of their lives. Mobile, for example, has a more than 10% share of adults' media time each day, but less than 1% of ad dollars. . . . On the flip side, newspapers and magazines continue to command ad dollars far ahead of their importance in consumers' day.[7]

This book has documented the failure of the newspaper industry to exploit the digital era to this point, but the forecast for significant growth in mobile applications and advertisers eyeing the burgeoning mobile audience offer newspaper companies an opportunity for digital redemption. Albarran wrote, "Today every media company has a mobile strategy, and recognizes this is another evolving market in which firms must be active in order to engage audiences."[8] *Editor & Publisher* proclaimed at the end of 2011 that the market's message had been received: "Print enterprises are well aware that the digital incursion has become a revolution, and as a result many have taken up the cry of 'digital first' as their business mantra."[9]

Echoing lessons that should have been learned from the industry's Internet experience, the *American Journalism Review* warned content publishers that "the tablet app format warrants a set of standards and practices of its own."[10] Yet an industry study found that many publishers were ignoring some popular conventions of tablet devices. For example, users expect applications to re-orient their content when the device is shifted from a vertical position to a horizontal position. "But because this takes resources, we're seeing a move toward fixed orientation," said the study's author.[11] Nevertheless, the *American Journalism Review* was willing to give tablet publishers a pass more than a year after Apple introduced its iPad, arguing that "criticism feels a little cheap at this stage of the game. The news organizations that are pioneering this space and providing lessons for others don't deserve to be judged for early missteps."[12]

Ironically, an article in the *Columbia Journalism Review* 20 years ago matter-of-factly reported that "Knight-Ridder is planning for the day when multi-media newspapers will be available on portable, touch-sensitive, flat-panel displays."[13] Now that such a day has arrived, Knight-Ridder is no longer around to participate in it. For newspaper companies still operating in 2012, however, Knight-Ridder's demise should serve as a reminder that little margin for error remains. Achieving success in the mobile marketplace is imperative if the newspaper industry has any chance to reverse the business trends of the previous disastrous decade.

The newspaper industry's financial condition deteriorated severely in the 2000s. Newspaper companies collected just $25.8 billion in advertising revenue in 2010, which was nearly a 50 percent decline from the industry's peak in 2005. Only 12 percent of the total advertising was attributed to online advertising.[14] This steep drop in advertising revenue and continuing circulation erosion resulted in closed newspapers, bankruptcy filings, and continued layoffs throughout the industry. The example of Knight-Ridder's demise was indeed a harbinger of the turmoil to come. More recently, however, Gannett's reorganization of *USA Today* combined with other activity throughout the industry—such as the *New York Times'* links with social networking sites like Facebook and Twitter and a preponderance of deals to distribute content through wireless applications—gave credence to analysts who maintained that "major changes" in the industry were underway.[15]

Many newspaper companies were busy erecting paywalls designed to charge readers for access to online and mobile content. *The Economist* suggested that paywalls were necessary, at least in part, because of the mobile push: "Many newspapers have created paid-for apps. There is little point doing that if a tablet user can simply read the news for free on a web browser."[16] Fewer than 30 online newspapers were ensconced behind a paywall in April 2010, but more than 100 were deploying them by late 2011.[17] Even so, financial analysts were not convinced the decisions to extract dollars in this manner would be enough to make any meaningful long-term economic difference for the industry. The *New York Times*, for example, reported in September 2011 that it had signed up 324,000 online users, a number that was lauded by many in the industry as a positive indicator of the potential for paywalls.[18] An analyst for Citigroup was not convinced those numbers mattered. He stated that even if those digital subscribers contributed as much as $75 million to the company's annual earnings, it would still fall at least $5 million short of offsetting the company's losses stemming from erosion in print advertising.[19]

In 2012, the erosion in print advertising continued, but no digital strategy had clearly formed to provide newspaper companies a safety net. For more than 30 years, newspaper companies had engaged in rhetoric defending their plans to conquer interactive media. At this juncture, the industry's survival required more than rhetoric emanating from tactics that were generating far less revenue than the industry was losing.

This chapter next presents a summation of the key themes that have emerged since 1980 when newspaper companies first began experimenting with online media. In doing so, it demonstrates persistence in the way newspaper companies responded to the constantly changing, threatening, but also enticing emergence of electronic information and communications networks. Persistence, however, did not translate into financial success. Therefore, this review of key themes provides insights into why newspaper companies failed to exploit online media to their advantage.

This chapter also revisits several terms presented in the Introduction: new media, interactivity, and convergence, and reflects on how the newspaper industry interpreted these terms in practical usage and the ramifications such interpretations had on the industry's actions. Together, these two sections address issues that relate to the newspaper industry's internal culture and how that culture influenced its business model. These issues should be considered anew as newspaper companies confront the technological shift into mobile media and contemplate actions that will determine whether or not they have a future.

Review of Key Themes

At the outset of the online era in 1980, newspaper companies were pioneers. The newspaper industry approached the potential of an online market with the confidence that accompanied its standing as a leading industrial employer and top advertising distributor. The industry's economic clout allowed for it to relish its social responsibilities as the premier defender of the rights to free speech and a free press. Newspaper executives had no fear of the online market in the early 1980s. They assumed their companies would adapt to new technology, continue their local market dominance, and honor their implied contract with the American people under the First Amendment.

Over time, however, the newspaper industry lost its leadership position. The information marketplace changed dramatically in a relatively short period, and the newspaper industry struggled to adapt. The history of this period reveals seven key themes that can be seen, retrospectively, as having significant influence on newspaper companies and their collective decision making as their industry confronted the online market from its inception. These themes, as much as the historical story itself, are of critical importance in understanding what might be happening now in online and mobile media, and in foreseeing or even shaping future events. These key themes fall largely into two broad categories: the organizational and professional culture of the newspaper companies, and the business models that sustained the newspaper industry. The two categories are inextricably linked as the culture of the newspaper industry profoundly influenced its approach to business.

Freedom of the Press and Its Influence

The first of the key themes to be addressed is instrumental in understanding how the newspaper industry collectively saw itself as being different and apart from other businesses. The newspaper industry historically viewed the First Amendment's constitutional guarantee of a free press as a social contract with the American public. Udell argued that newspaper companies indeed occupy a unique position in American industry because they are dependent on the free enterprise system for their livelihood, but are protected by the free press provisions of the U.S. Constitution.[20] This status separated newspaper companies—in the view of their owners and employees—from other business and contributed to a cultural sense that newspapers represented a national institution sustained by private enterprise, but in business to serve a public need. For decades, newspaper publishers used their free speech responsibilities as justification for their large profit margins. Newspaper companies had to be exceptionally profitable, according to this position, so that they would not be beholden to any single advertiser or other outside party. Exceptional profits were believed to give newspaper companies the independence required to uphold their end of the social contract free of influence.

From the very outset of the electronic era, newspaper professionals were concerned that electronic distribution would alter their relationship with government authority. Newspaper publishers saw how their electronic rivals in television and radio had succumbed to intense regulatory control by the FCC because they relied on the public airwaves to distribute their content. Many newspaper people believed that distributing their content over regulated telephone wires would make them susceptible to government oversight and weaken the protection afforded by the First Amendment.

This perspective weakened as electronic distribution grew more prevalent, and the newspaper industry turned the regulatory argument in a different direction when it lobbied legislators and regulators to limit the role the telecommunications industry could play in the distribution of electronic information (see Chapter 3). Newspaper leaders argued that the telephone industry—originally a giant monopoly and later a collection of regional operating companies—was so powerful that its unfettered entry into the market would force others out, thereby reducing the diversity of information sources. Although the diversity argument swayed politicians and judges, mainstream consumers perceived newspapers as obstructing technological progress. Even some critics from within believed the newspaper industry's argument ran counter to its values. These critics believed that newspapers should have championed the telecommunication industry's right to free speech rather than using it against them.

More recently, the financial plight of newspapers has caused some to question the viability of the larger concept of journalism. This notion underscores how closely journalism has been associated with printed newspapers; if newspapers are failing, then journalism must also be in peril. There are industry executives who now argue that journalism as content must be compelling enough to transcend its product form.[21] However, the newspaper industry has been defined by its journalism-centered culture, which historically has been reluctant to embrace change that threatened its print heritage.

The Pre-eminence of Print

The newspaper industry's insistence on keeping the printed newspaper form at the forefront of its product offerings is a key theme that explains the industry's relationship with online media: printed newspapers were always the primary product focus, while online endeavors were relegated to ancillary status. Chapter 5 noted how the advent of online distribution, especially the emergence of the web, should have forced the industry to assess the underlying fundamental nature of its enterprise: were newspaper companies in the business of printing and distributing newspapers, or were they in the business of supplying news and information? How newspaper companies answered this question was essential to facilitating the necessary shift in business models.

Jones explained that printed media such as newspapers "sold space," which he explained as "trading on the attention people would pay to the spatial organization of the printed—mediated—word."[22] He contrasted that model with "electronic broadcast media," writing that such media "sold time," or the amounts of it that someone would spend with "the temporal organization of radio and TV."[23] He wrote that Internet-based media, however, "sell attention, without regard to space and time," suggesting that this model values "connection and linking."[24] The newspaper industry's approach ignored this fundamental shift in how audiences were valued in the online market.

Throughout the digital era, the newspaper industry had many opportunities to adjust, to shift its focus and its economic dependence away from print; but the industry never fully committed to an electronic future.

Throughout the online era, newspaper companies were told that they should lessen their dependence on print and reduce the expensive distribution overhead associated with it. At the outset, newspaper companies were enamored with the potential cost savings of digital delivery, but as recounted in Chapter 1, the enthusiasm for those projects waned when consumers gravitated to interactive functionality rather than newspaper content. Also in the early years of the digital era, the newspaper industry invested heavily in news delivery over cable television systems. Newspaper companies abandoned those projects, however, when they recognized that consumers had little interest in watching passive displays of text on television (see Chapter 2). The point behind both of those examples is that newspaper companies were looking to replicate their printed newspaper in digital form. When that did not materialize, newspaper companies withdrew.

The pre-eminence afforded the printed newspaper was especially acute in the early Internet period. Newspaper companies developed automated programs to transfer content from the printed product directly to the web, but invested little in original online content. Furthermore, newspaper companies in those early years refused to allow news to appear online before it was published in print, fearing their own internal competition (see Chapter 4). The result was a lukewarm entry into a red-hot market that relegated the newspaper industry to a laggard position from which it never recovered. As one exasperated commentator observed: the newspaper industry must "stop worrying about how the news is delivered."[25] When challenged to decide if they were in the business of selling newspapers or, more holistically, information, newspaper companies should have treated the question as more than a rhetorical exercise.

A Culture of Information Control

The newspaper industry's desire to maintain its print-centric focus has roots in another key theme discerned from the history. This theme—control of news and information—was so ingrained in newspaper culture that it transcended the business model; it was a tenet of journalism doctrine manifested in the newspaper's agenda-setting role (see Chapter 5). Most newspaper editors and executives took their journalistic responsibility seriously as integral to the country's democracy. As noted above, newspaper people saw their industry as "a unique social institution" within the political system and believed they were fulfilling a social contract necessary for democracy to operate.[26] However, newspaper companies operated from the premise that it was their job to define the news, the degree of its importance, and when that information would be disseminated. The entire newspaper process was rooted in control. Reporters had access to information, sources, and events the general public did not; editors determined what events would be featured and how the stories would be written and presented; the paper was printed and distributed in its entirety without regard for individual preferences. A consumer received the sports section, for example, even if that reader did not follow sports. A consumer held the choice whether to subscribe, but other feedback was limited to a letter to the editor, and the newspaper controlled which of those it would print. As Schonfeld observed, "historically, the most successful media companies have controlled both content and distribution."[27] Until the online era began, this model operated mostly unchecked for more than two centuries.

To an extent, radio and television already had disrupted the dissemination process. Online media, however, was entirely different because it provided newspaper readers their own media platform outside of the agenda-setting reach of organized newsrooms. Bryant wrote that the extensive choice associated with the Information Age changed the relationship dynamic that existed between media and audience, creating what he described as the "sovereign consumer."[28] Newspaper people reacted poorly to this trend. Rather than recognize an opportunity to organize community and foster discussion, newspapers acted as if their social contract had been violated. Journalists especially detested the notion of giving up control demanded by the Internet, and they were not particularly enamored with having to interact with readers, either.

Perhaps the newspaper industry's most blatant attempt to exert control over the emerging Internet occurred with the inception of NCN (see Chapter 4). As a case study, NCN is remembered as a significant failure because of infighting among the newspaper companies that owned it and their inability to find an effective business model for it. NCN was, in effect, an online portal for the newspaper industry's entire collection of content before the portal construct actually emerged. From the moment it was created, however, most of NCN's owners saw it as an enforcement device: an entity that could marshal the collective weight of the newspaper industry to bring order to the Internet's perceived chaos. The newspaper companies believed that through NCN, they could influence everything from advertising sizes to browser compatibility issues. It was an impossible mission to fulfill, but the failure of NCN did not dissuade others in the industry from believing that the Internet could be controlled.

As noted above, newspaper companies were reluctant to distribute their content online before it had first appeared in print out of fear the overall franchise would be weakened through self-competition. But the culture of control manifested itself in other ways as well. When newspapers offered interactive functions such as message boards, for example, they were mostly moderated and censored. Even when printed newspapers included email addresses for readers to contact reporters and editors, the public inquiries that ensued went largely unanswered.

The prospect of a newspaper facilitating discussion rather than controlling it was the antithesis of newspaper doctrine. Moreover, the repeated attempts to exert control over what was essentially an open platform kept newspaper companies from fully accepting the Internet as an inherently different medium. As such, newspaper companies kept their online editions subservient to their print editions. The irony of this approach is that the more newspaper companies gave in to their culture of control, the more control they seemed to lose.

Culture Trumps Innovation

As discussed in the opening of this section, newspaper companies began the online era as pioneers. The development of the early videotext projects and the investments in cable television systems (as explored in Chapters 1 and 2, respectively) were extremely innovative in the early 1980s. However, a key theme that emerged from this history is that this early innovative spirit was quashed by a culture that was inherently risk averse. Throughout the past three decades, newspaper executives intellectually acknowledged that their industry needed to accept change in order to survive in the digital era. In the era described in Chapter 1, for example, the newspaper industry was at the forefront of pioneering online delivery with videotext projects. Chapter 4 recalls the ambitious attempt by the newspaper industry to fund a cooperative research and development effort based at the Massachusetts Institute of Technology, and Chapter 5 recounts the example of NCN as the newspaper industry's initial aggressive push to exert its influence in the emerging Internet era.

Emotionally, however, the industry never embraced the depth of change required to prevent the economic calamity that led to the industry's decimated condition. Observers point to the underlying conservative nature of the industry's business practices as fostering complacency. In Chapter 6, the discussion of operational convergence, for example, noted how employees interested in enabling change—internal "entrepreneurs," as they were described—were often assigned to online departments but afforded no real authority to implement policies that would have affected their larger organizations. Sequestering entrepreneurs, however, was only symptomatic of larger cultural issues that affected how the newspaper industry reacted to the challenges of online media.

As the newspaper industry confronted a series of unforeseen shifts in technology, it grew increasingly uncomfortable with its surroundings. The newspaper industry entered the online market in the early 1980s from a position of power and influence, but the surprising surge of cable television and the power plays of the telecommunications industry forced the newspaper industry into a defensive posture from which it never recovered. After playing offense for several decades following the Great Depression of the 1930s, the newspaper industry found itself trying to protect its markets rather than expand beyond them.

Operating from this defensive position, newspaper companies essentially eschewed innovation. As discussed in Chapter 3, newspaper industry trade groups lobbied lawmakers and regulators to restrict the telecommunications industry's ability to distribute electronic information. While those efforts were largely successful, the newspaper industry suffered from a protectionist stance that caused it to fall behind more technically savvy competitors. The efforts to fund research and development such as the one discussed in Chapter 4 turned out to be too limited to affect the market. Rather than invest in new technologies at a level required to make a difference, newspaper companies allowed those funds to flow to the bottom line, inflating current profits at the expense of long-term planning. Chapter 4 discussesd the newspaper industry's apparent dismissal of innovation as the result of a short-term focus inherent in an industry required to produce an entirely new product every day. Although this was referred to as a pragmatic approach, the culture it fostered led to risk management strategies that embraced partnerships and alliances instead of direct industry investment in innovation. While this strategy worked well in terms of preserving capital, it meant the newspaper industry was relegated to a role of content provider on someone else's platform. When the market turned dramatically toward the Internet, the newspaper industry was unprepared to stand on its own in a marketplace driven by innovation.

The Lasting Effects of Early Online Failure

The newspaper industry's dismissive attitude about innovation had roots in the failure of some of its earliest online efforts. One of the key themes that emerged from this retrospective study is that the early videotext projects had negative, long-term implications for the newspaper industry's approach to online media because of their characterization as failures. Newspaper executives had been heavily influenced by the prevailing Information Society rhetoric and, as shown in Chapter 1, moved to implement technology that would transform the rhetoric into reality. However, the early videotext technology did not appeal to consumers, and newspaper projects built on that platform never achieved any audience traction. When Knight-Ridder closed its Viewtron project, followed closely by Times Mirror's decision to shut down Gateway, the industry was chastened. There had been so much hyperbole surrounding these services and the role newspaper companies would play in ushering in a new era, the failure of these projects left a scar on the industry that never healed. More than a decade later, newspaper executives were still pointing to Viewtron and Gateway as reasons why the industry should approach electronic services with extreme caution (see Chapter 4).

By allowing projects such as these to be branded as failures rather than as risk-taking research and development, the industry began a descent into technology cynicism that fomented protectionism; this was examined in the Chapter 3 discussion of the newspaper industry's strident opposition to the telecommunications industry's entry into electronic information delivery. Newspaper companies that were heralded as technology pioneers in the early 1980s were by the end of the decade seen in a completely different light. Had the newspaper industry considered its early videotext projects in the spirit of innovation and experimental learning rather than as a means to an end, the setup to the Internet era likely would have been very different.

The early online projects demonstrated that consumers wanted more than a passive experience. When content was distributed through a networked computer, the natural instinct for a consumer was a desire to do something with it. Consumers wanted to respond, to share content with others, to alter it in ways that would give them ownership. In other words, the early projects pointed to the importance of interactivity as an online attribute. From the outset, however, newspaper companies viewed their content as sacrosanct. In protecting what they viewed as an asset, newspaper companies used their online platforms to create electronic versions of printed newspapers. In the process, they ignored how consumers wanted to use online information. In recalling the Viewtron project, for example, Boczkowski found that Knight-Ridder "neglected its own usage data showing that adopters were more interested in communicating with each other than reading newsroom-generated content."[29] Chapter 1 recalled how the discovery of this consumer behavior contributed to the company's decision to end the project. When consumers did not respond to Viewtron as a newspaper, the newspaper company that created it was not inclined to support its growth and development.

This is the first key example of a time when newspapers largely missed the significance of a turning point in the evolution of media, and stubbornly clung to their print-centric publishing model. When consumers did not regard Viewtron or Gateway as newspapers per se, the newspaper industry shut them down rather than try to respond to the interactive services consumers wanted in an online experience. When the online market moved into the proprietary services era as recounted in Chapter 4, newspapers failed to grasp that consumers were turning to these services largely for their email and messaging platforms. Content was ancillary, but newspaper companies partnered with AOL and Prodigy because they afforded publishers some semblance of control. The proprietary model allowed newspaper companies the ability to organize and distribute their work supported by advertising. The close resemblance to the offline publishing model appealed to newspaper publishers, but as discussed in Chapter 4, the resulting gated communities (content on Prodigy was not available to subscribers of AOL and vice versa) left these companies exposed when the open network of the Internet arrived.

The Internet did not conform to any control model, which explains why it represented such a disruptive technological shift for the newspaper industry. With the Internet, consumers could take control of content; they no longer needed a publisher's consent. As explored in Chapters 5 and 6, this fundamental difference in the Internet versus previous online platforms was ignored by newspaper companies at the outset and represented another period when they missed a significant turning point in the evolution of media technology.

The newspaper industry's reaction to the Internet can now be seen as also having roots in issues discussed earlier, such as information control and a reliance on the print model. The outcome of the industry's activities clearly indicates that they were an insufficient response to meet the competitive demands of the marketplace, especially in protecting vital revenue streams.

Advertising Position Left Unprotected

Throughout the digital era, the newspaper industry was warned that online distribution threatened its advertising revenue. However, a key theme that emerged from the history was the newspaper industry's failure to heed those warnings, thereby leaving its advertising position unprotected against many new competitors. The newspaper industry's actions in the early part of the digital era—specifically investments in cable television (see Chapter 2), the protracted lobbying battle against the telecommunications industry (see Chapter 3), and alliances with proprietary online services (see Chapter 4)—were largely viewed as attempts to defend its advertising position. In the case of cable television, for example, newspapers saw cable franchises as direct competition in local markets. The newspaper industry's lobbying to keep AT&T and the telecommunications industry out of the online information business—although framed as necessary to protect information diversity—was also aimed at stalling online competition with vast databases of Yellow Pages advertising. During the brief period when newspapers formed alliances with proprietary services, Prodigy emerged as the industry's most-preferred partner because its advertising model was seen to be the most closely related to the newspaper industry's advertising model.

With so much attention given to advertising concerns during the pre-Internet portion of the online era, it is perplexing to see how vulnerable newspaper companies were when upstart companies targeted their markets as the Internet emerged. Google's search-based advertising, eBay's auction listings, and Monster.com's online job postings are examples of online advertising that contributed to significant erosion in the newspaper advertising market, especially in the classified advertising category. The newspaper industry's failure to protect its online market for classified advertising is especially problematic given the early and frequent warnings that this category of revenue was perhaps the most vulnerable to online competition.

Chapter 1, for example, presented Compaine's observation from 1980 that "classified advertising . . . does lend itself more to these futuristic delivery modes," adding that "this would appear to be one area in which newspapers may well have to take the lead, before . . . others usurp this function."[30] As noted in Chapter 6, however, newspaper companies did not take the lead in developing online classi-fied technologies and trailed their competition as the online era transformed into the Internet era. Newspaper companies placed their classified advertising online through platforms that were not as interactive or sophisticated as the competition. The industry tried to create scale and share risk through numerous joint ventures and investments in new companies, but had little success with those efforts. From the perspective of industry analysts and consultants, the newspaper industry's approach to advertising—especially classified advertising—was reflective of the overall pattern of operating without a coherent strategy. Chapter 6 features a com-ment from the chief executive officer of Times Mirror in 2000 when he acknowl-edged that the newspaper industry had taken classified advertising for granted. Newspaper company managers had allowed such complacency to control their industry's destiny. These executives had dismissed the need for innovation, and the results underpin this key theme. The advertising position was exposed to an onslaught of new competitors borne out of the very innovations and inventions newspaper companies chose not to exploit.

The complacency that underlies such decisions had its roots in profitability. The following material examines how the relentless pursuit of profits ultimately created an industry full of short-sighted companies unprepared for dealing with the rapid shifting technologies of the digital era.

Emphasis on Short-Term Profitability over Long-Term Planning

The final key theme identified in this analysis—an emphasis on short-term profit-ability over long-term planning—can be seen as the unifying idea of this book. The newspaper industry's pursuit of profits led to decisions that were detrimental to the long-term health of the industry. In Chapter 6, an industry observer noted the emotional issues newspaper executives had in accepting fundamental shifts in media revenue. He stated: "It's really hard to cannibalize yourself and trade high-margin revenues for low-margin revenues one second before you have to."[31] Delaying that transition, however, proved to be a dangerous tactic as the industry mistimed its response. The market for printed newspapers deteriorated faster than publishers anticipated, and they had not developed their digital operations to the extent that online revenue could offset the declines in print revenue. Throughout the 2000s, newspaper companies saw their Internet operations as supplemental to the core printed product and managed them as an expense to be controlled. Newspaper executives never embraced their online editions as engines of growth worthy of long-term investment.

By the time newspaper executives understood their new reality, it was too late to make up the lost ground and affect their immediate economic circumstances. In Chapter 6, Rosen is quoted as describing this as a period of "profitable demise," underscoring the notion that newspaper companies had maintained high profits only by jeopardizing long-term viability.[32] Within this context, Knight-Ridder's decision to sell its newspapers and shut down, also presented in Chapter 6, can be viewed as the symbolic end of the traditional newspaper model. As such, the newspaper industry continued to deteriorate in the latter half of the 2000s. In many cases, those coveted profit margins turned to losses, and the few remaining investors bid down the value of the stocks. Some newspaper companies lost 80 percent of their market value, and the entire sector traded on financial markets at or near historic lows by the end of the decade.[33]

Some would argue that the condition of the newspaper industry in 2012 should be expected of any mature industry at the end of a natural life cycle. In the case of the newspaper industry, however, its history presents a compelling argument against natural evolutionary forces. As a Wharton Business School report asserted: "newspapers themselves are to blame for a large part of the problem,"[34] indicating that a willingness to adopt a more long-term focus and make different strategic decisions could have altered the industry's circumstances.

Chapter 1 recounted how the newspaper industry cast the early videotext projects as failures and fixated on how much money had been lost. The industry could have cast that money as an investment in research and development. Chapter 3 chronicleed a decade of the newspaper industry waging a protectionist campaign against the telecommunications industry rather than investing in its own technology innovation. Chapter 5 discussed the closing of NCN as the failure of leading newspaper companies to cooperate. Imagine the altered landscape if NCN had emerged as a well-funded portal rather than disintegrating amidst industry infighting.

As the history demonstrated, U.S. newspaper companies faced many decisions during the online era where conservative responses assigned the industry to its predicament. With the collapse of advertising revenue in its print products, the industry's old business model has been discredited. But a completely new model has yet to emerge, especially given that newspapers' online advertising revenue has yet to come close to making up the print advertising shortfall. Through employee layoffs and other budget cuts, most newspaper companies had by 2012 stemmed the financial losses that had crippled them in the latter half of the previous decade. But even the most ardent supporters of the newspaper industry understand that media economics make it highly unlikely that newspaper companies will ever again be able to rely on printed products for long-term growth and sustainability.

Newspaper companies now speak of their operations as multiplatform, and the Gannett example of the changes at *USA Today* that opened this chapter could signal an effort at real operating reforms. Once again, the future of the newspaper industry rests on its ability to embrace *new media* products that are innovative and accept *interactivity* as a core attribute as media *convergence* becomes increasingly about devices. The intentional emphasis included in the previous sentence illustrates that the three key concepts presented in the introduction as a foundation for this book continue to be relevant as the newspaper industry contemplates a future dominated by mobile devices. The following section reviews these concepts and suggests that understanding the newspaper industry's relationship with them helps to explain how the industry must adapt going forward.

Concepts Revisited

This book presented three key terms in the Introduction that related to the development of online media, including new media and the concepts of interactivity and convergence. How newspaper executives came to understand what these terms represented for their own industry is reflected in their responses to development of online media. In relating to new media, for example, newspaper executives accepted "new" to mean online media and digital media. They did not appreciate, however, that when applied to Internet-based media, "new" truly represented a different paradigm that required new modes for operating. Each of these three key terms is revisited in light of the history presented to more fully discuss their significance.

"New Media" Misunderstood

Newspaper executives believed from the outset that a key strength of their industry rested in its ability to organize information. It was expected that newspaper companies would translate those organization skills to their online platforms. As quoted in Chapter 5, an industry leader stated: "It is information *processing* that is newspapers' greatest strength," adding that "newspapers must understand that what they do best—gathering, packaging and distributing news and information—is much more than ink on paper."[35] Newspaper executives clearly understood that electronic distribution represented a physical change in product form from paper to digital, but this understanding represented the limits of how newspaper executives defined "new media."

For newspaper executives, the term "new media" was synonymous with online distribution. But that simple definition, while not entirely incorrect, did not allow them to frame the broader implications that online distribution represented. They did not grasp that all online distribution would not be the same. As such, they did not recognize that Internet-based new media represented a new paradigm that destroyed the structure of content as they knew it.

Newspaper formats belonged to an era of linear information processing in the McLuhan vernacular[36] and relied on the hierarchal concept of presenting information from most important to least important. The advent of hypertext allowed users to approach information from a multidimensional perspective. Readers could access information at any point in the continuum and could navigate from one point to another based on their own interests and desires. Furthermore, online users could choose content without advertising or advertising without content. And in some cases—Monster.com and eBay are examples—the advertising was the content. In Chapter 5, an industry analyst is quoted: "the components of what we historically know as the newspaper have become unbundled."[37] Allen referred to this phenomenon as the emergence of "not media" when he explained that companies such as Google do not operate in the ways normally associated with media, but merely "trespass into the economic fiefdoms of media."[38]

Newspaper people had no response for the unbundling process and the ensuing creations that did not fit neatly into their media definition. The newspaper industry's perspective relied on the tenet of content creation: that consumers responded to original content and in turn created a healthy information exchange conducive to selling advertisement. In reality, however, consumers gave less credence to the original content of newspapers than executives and journalists were willing to believe. This became especially problematic for newspapers as the abundance of news available electronically gave such content the characteristics of a commodity and reduced its perceived value. Even so, consumers welcomed structure, and they rewarded those companies that provided it in ways that enhanced rather than dictated an information experience.

Yahoo and Google are examples of innovators that filled the structural void of the Internet with portals and robust search engines—models that relied on content aggregation rather than content creation. As quoted in Chapter 5, the chief executive officer of Yahoo stated: "I don't think old media is what people are going to spend most of their time doing on the Internet. This paradigm needs its own inventions, its own methods, its own way to go forward."[39] In that sense, "new media" became identified with breaking media tradition by enabling and empowering interactivity.

Interactivity Defines the Paradigm Shift

From the outset of their earliest online endeavors, newspaper companies learned about the inherent interactive nature of online media. However, newspaper companies mostly ignored the interactive capabilities of online systems. They used online as a form of distribution to push out news and information to consumers, but it was rarely considered as an opportunity for two-way communication with readers. Some interactive functionality was tolerated within the walled gardens of the proprietary systems such as Prodigy and AOL as discussed in Chapter 4, but newspaper companies reverted to their early videotext mentality when migrating to the Internet in the mid-1990s. There was gradual acceptance of interactive features throughout the first decade of the Internet, and online editions eventually featured full complements of interactive functionality such as blogs and message boards. But critics contended that newspaper companies never appreciated interactivity to the point that would cause them to make interactivity the focal point of how their online editions operated.

Such criticism reflected the frustration of many industry commentators who saw the newspaper industry's failure to embrace interactive functions as short sighted. When newspaper companies initially automated content repurposing, the practice was accepted as necessary for expediency to market. As the practice became commonplace, however, critics contended that newspaper companies were taking an easy route to the Internet, but one that failed to take advantage of either immediacy or interactivity.

As the Internet developed, interactivity emerged as the major reason a media paradigm shift occurred. The Internet was ideally suited for interactivity because its creation was rooted in the development of a communication platform that was deliberately designed to facilitate sharing of data and ideas. Entrepreneurs who were not wedded to the artifacts of media tradition embraced the Internet's open platform and introduced new media products that empowered consumers. Newspaper companies did not understand what was taking place until after fully developed competition appeared. Newspaper companies initially saw the Internet as just another wire and were skeptical of its long-term viability given that it had no ownership or governing body in the traditional sense. Its interactive nature, however, had, in effect, made all of its users de facto owners, which was a concept the newspaper industry did not appreciate. Furthermore, newspaper companies ignored interactivity because embracing it would have meant ceding control. Culturally, newspapers were aloof and detached from their readers. Interactivity, therefore, represented the antithesis of newspaper culture.

The reluctance of newspaper people to appreciate the Internet's inherent interactivity was manifested in the tepid response mounted against Internet-enabled competitors. Whether or not this grew out of a risk-averse business culture or a culture steeped in agenda-setting journalism can be debated, but the history demonstrates that newspaper companies only reacted to online media out of a perceived necessity. They never fully committed their operations to online media, and never embraced online media as their core business. It is difficult to win a war when the troops are not convinced of the cause. In the case of the Internet, the newspaper industry simply did not believe in interactivity, and therefore, it could not muster the passion necessary to compete.

If interactivity can be seen as the essence of online media, as this line of thinking indicates, then convergence can be used to describe the media evolution taking place around it. The many iterations of convergence have been discussed throughout this book. The section that follows discusses various perspectives of convergence and how the concept relates to the newspaper industry's relationship with online media.

Convergence: A Matter of Perspective

Newspaper companies initially were fearful that electronic distribution would allow others to become newspapers. The early videotext experiments were seen, at least in part, as a response to commercial services such as CompuServe that were entering the consumer market. The newspaper industry's political war with the telephone companies throughout much of the 1980s and early 1990s was aimed at preventing them from creating local news and information services. Later, as dozens of newspaper companies partnered with Prodigy, they bargained with the proprietary service to eliminate its own content staff and drop plans to create original content.

Newspaper companies approached the advent of electronic distribution from an industry-centric point of view. As such, the concept of convergence was considered rather narrowly. It was a technological construct that defined the process of computers connecting to the telephone infrastructure for the purpose of delivering information. Newspaper companies, therefore, understood convergence as a technology construct separate from a media construct. In this context, convergence was about enabling technology that would allow newspaper companies a new way to distribute their content.

As noted in the Introduction, there is wide acceptance among scholars that convergence, as it is discussed in the context of new media, began as technology shifted from analog to digital. As this shift transpired, newspaper companies approached convergence as principally about enabling new production and distribution methods. As a result, they spent much of the online era looking at the media landscape through the lenses of mergers and acquisitions or strategic alliances and partnerships. As explained in the Introduction, Dominick referred to the former as "corporate convergence," or what happened when a single corporate entity would form to offer multiple media products, and the latter as "operational convergence," which occurred when competitive media outlets set aside their rivalry to cooperate on content creation, often with the intent of jointly producing material for online distribution.[40]

As newspapers considered convergence in these pragmatic terms, they formed partnerships with television stations, as explored through the example of Media General's efforts in Tampa (see Chapter 6). In the end, however, these types of initiatives were convergence in name only because they did not truly address the issue of creating content suited for the interactivity of new media. A newspaper and television station cooperating to produce content for a website likely saved both parties money in terms of production, but it was not convergence in the broader spirit of contributing to the evolution of media by creating something new out of old parts. Essentially, these projects turned out to be nothing more than distractions. They gave the industry the allusion of pursuing productive courses of action, but in reality the partnerships and alliances allowed the industry to avoid engaging with the larger, fundamental shift that the Internet represented.

Conclusion

The Internet often is described as a disruptive technology. Many enterprises across all sectors of business and society, from telecommunications providers to travel agencies, from software developers to politicians, have been transformed since the Internet emerged as a platform for citizens to communicate, to consume and produce media content, to conduct transactions, and to share information about themselves with others online. Yet the media has probably been more affected by (and also most challenged by) the Internet. This book has provided a detailed examination of how a specific medium—newspaper publishing in the United States—attempted to respond to the Internet, and in doing so provides a deeper understanding of the condition of the newspaper industry itself. Perhaps more significantly, however, this analysis of history reveals that the disruption for the newspaper industry had as much to do with its *understanding of* and *reactions to* the Internet as with the actual technology itself. All technology has been shown, inevitably, to be part of and not distinct from the human societies from which it emerges. The Internet is no different in this regard.

Clearly, the newspaper industry has been transformed by technology—any industry that loses nearly half of its primary revenue in less than a decade would find its business model discredited and its survivability questioned. In this book I have shown that the newspaper industry was not a passive bystander as Internet-based competition invaded its markets. It would be incorrect to characterize the newspaper industry as a victim of technology, that the disruption and subsequent transformation of the newspaper industry was entirely out of its control. Although the newspaper industry had no influence on the timing of the Internet's emergence, it was in complete control of its reaction to it.

The newspaper industry had ample warning about the shifts occurring in its market. Newspaper companies learned early in the digital era about the vulnerability of their classified advertising franchise, yet failed to protect it. Newspaper companies learned about the significance of interactivity, yet ignored their audience's pleas for a relationship. When a business knows of its weaknesses and realizes that its marketplace is changing, but does not have the conviction to address its problems forthrightly, its deteriorated condition cannot be blamed on technological disruption. The newspaper industry arrived at its current predicament on its own volition through the diluted choices made and the lackluster competitive response that ensued.

This book began by exploring the advent of the online era and the creation of the first online newspapers in the early 1980s. It is important to consider this earlier perspective because the newspaper industry in the 1990s was not reacting to the Internet in a vacuum. From the vantage point of the newspaper industry, the Internet was part of a technological continuum that began with videotext and progressed through the period of proprietary online systems. These earlier platforms had served to inform newspaper companies that online systems were first and foremost communication platforms. Users of such systems felt empowered by a level of control that was missing in the realm of traditional media. This user empowerment was apparent even in the systems that preceded the Internet. However, even during that time, the newspaper industry ignored this reality, as they did again, more detrimentally, in the Internet era.

While the Internet may have been part of a continuum in online technology, its arrival—more specifically, the arrival of the content presentation layer known as the World Wide Web—represented a radical departure from earlier online systems. The Internet was far more decentralized, being an end-to-end network in which each computer attached to the Internet might be both a client and a server, and whose operations were governed by protocols that deliberately encouraged open interconnection rather than proprietary closure. The early online models were similar to publishing models with content created centrally and distributed to subscribers. Newspaper companies completely misunderstood the openness inherent in the Internet and stubbornly tried to force a control-based publishing model into a platform that fundamentally was designed to operate without central control. Ignoring the audience's desire for interactivity and empowerment in 1985 led to a series of small online failures; ignoring those same wishes in 1995 created the conditions for the newspaper industry's cataclysmic failing.

Nevertheless, a smaller and retrenched industry has survived to face yet another period where strategic decisions must be made in regards to a media industry inflection point: the emergence of wireless computing and mobile media, including smartphones and tablet devices. However chastened the newspaper industry may be in the wake of its recent history, it has an opportunity to be resurrected as a content provider for these new platforms. Doing so, however, will require that newspaper companies accept that the media landscape has changed inalterably. Once that fundamental concession is made, newspaper companies must then adapt in ways that have eluded them previously.

Although many activities were undertaken by newspaper companies regarding online media, much of the effort was a fight against the natural forces of convergence that throughout history have allowed new forms of media to supplant older forms. Even though newspaper companies created online media since the early 1980s, they never embraced the natural evolutionary process and refused to free their online products from the artifacts of a print heritage. Mostly, they rejected interactivity, which emerged as the defining attribute of online media. In doing so, the redemptive power of convergence was lost and newspaper companies ended up losing the one attribute that had mattered to them: control of the information they created and disseminated.

If newspaper companies are to find new life within the ecosystem of applications fostered by wireless devices, they will have to address deep-seated industry cultural issues that led to their current predicament. Newspaper companies must be willing to acknowledge that their primary product can no longer be a printed edition. The demise of Knight-Ridder Inc. in 2005 was the alarm sounding that the traditional business model for newspaper companies was finished. However, it took another few years of economic failure, made worse by a deep economic recession, for the majority of industry leaders to see Knight-Ridder's closure as the bellwether it was. The Gannett restructuring example that provided an opening for Chapter 7 is an important, tangible sign of change. The announcement from an industry leader that print was no longer the primary product for its *USA Today* flagship newspaper was a radical admission. Given that newspaper companies have been so reluctant to renounce their print heritage, Gannett's announcement could emerge as the cathartic first step necessary to spur the overall industry's resurgence.

For newspaper companies to revive their economic fortunes, however, there must be more than announcements; a cultural transformation that replaces naysayers with innovators and risk-averse managers with those willing to abandon convention must occur. The newspaper industry can no longer afford to talk about change; it must embody change. This requirement is deep cultural change, not the type of change that resides only at the organizational surface that was brought on by downsizing and budget cuts.

Newspapers have a long history of experimentation. The early videotext projects, alliances with Prodigy, the formation of NCN and the creation of a 1990s-era prototype tablet are examples where the industry showed its ability to think about its future in new ways. In these and other examples, however, the newspaper industry failed to convert experimentation into innovation. In the final analysis of these milestone experiments, the newspaper industry's conservative culture blocked new approaches from achieving significance.

The newspaper industry first engaged in new media activity because it felt an obligation to deliver the Information Society before someone else could. The issue, however, quickly devolved into defining what the Information Society rhetoric meant in terms of practical product development. As the concept evolved to encompass much more than news and information, the gap between this market reality and the newspaper industry's willingness to adapt grew increasingly wider. By the late 2000s, the newspaper industry had accepted—albeit stubbornly— some of the tenets of the online era, and provided functionality that should have been present all along. In any case, it was too late for those concessions to make any significant difference in the era of the World Wide Web. The result in 2012 is a newspaper industry that is a mere shadow of the American institution it was at the threshold of the digital era in 1980.

Nevertheless, the industry has a chance to rebound as technology shifts again. The U.S. newspaper industry last faced such dire circumstances more than 70 years ago. Radio had threatened as a new media phenomenon in the late 1920s, and the Great Depression of the 1930s ravaged the newspaper industry. The newspaper industry was rescued in the 1940s in part due to citizens who wanted all the information they could get regarding World War II. Further bolstered by post-war economic expansion, the newspaper industry rebounded from the Depression era and prospered for four decades. The contemporary economy is far more complex than it was in the 1940s, and the roster of new media companies represents more formidable competition than radio did. Nevertheless, history suggests that it is possible for a turnaround to happen. Newspaper companies could find an economic model or a combination of several economic models to reverse their decline. Those models could be found in new forms of paid content or in new applications residing on smartphones and tablets. It is highly unlikely, however, that a rebound will occur in print. The future of newspaper companies rests now—just as it did in 1995—with the degree to which they are willing to separate from the printed paper and embrace new content distribution models.

Notes

Introduction

1. Littlewood, "A View from 88 Years Ago of Newspapers in Year 2000," 62. The 1903 article was attributed to T. Barron Russell, an editor at the *Pittsburgh Gazette*.
2. Ibid.
3. Newspaper Association of America (NAA), "Trends and Numbers: Advertising Expenditures." http://www.naa.org. NAA was formed on June 1, 1992, by the merger of seven associations serving the newspaper industry. The associations included the American Newspaper Publishers Association, the Newspaper Advertising Bureau, the Association of Newspaper Classified Advertising Managers, the International Circulation Managers Association, the International Newspapers Advertising and Marketing Executives, the Newspaper Advertising Co-op Network, and the Newspaper Research Council.
4. Sparks, *Media Effects Research*, 243.
5. McLuhan, *Understanding Media: The Extensions of Man*.
6. Hodge, "How the Medium Is the Message in the Unconscious of 'America Online,'" 341.
7. Sparks, 237.
8. Hodge, "How the Medium Is the Message," 341.
9. Ibid., 345.
10. Brooke, "Cybercommunities and McLuhan: A Retrospect," 23–24.

11. McLuhan, as quoted by Brooke, "Cybercommunities and McLuhan," 23.
12. Peters, "And Lead Us Not into Thinking the New Is New: A Bibliographic Case for New Media History," 15.
13. Kyrish, "From Videotext to the Internet: Lessons from Online Services 1981–1996," 5.
14. Ibid.
15. Stöber, "What Media Evolution Is: A Theoretical Approach to the History of New Media," 484.
16. Kaletsky, *Capitalism 4.0: The Birth of a New Economy in the Aftermath of Crisis*, 122.
17. Peters, "And Lead Us Not into Thinking the New Is New," 14.
18. Ibid.
19. Gitelman, *Always Already New: Media, History and the Data of Culture*, 1.
20. Holt and Perren, *Media Industries: History, Theory, and Method*, 2.
21. Winseck, "Back to the Future: Telecommunications, Online Information Services and Convergence from 1840 to 1910," 137.
22. Ibid., 140.
23. Ibid., 153.
24. Light, "Facsimile: A Forgotten 'New Medium' from the 20th Century," 371.
25. Gitleman, *Always Already New*, 1.
26. Ibid.
27. Holt and Perren, *Media Industries*, 1.
28. Ibid.
29. Boczkowski, *Digitizing the News: Innovation in Online Newspapers*, 50.
30. Smith, "Transition to Electronics: From a Bright Past to an Uncertain Future," 11.
31. Newspaper Association of America, "Trends and Numbers: Advertising Expenditures."
32. Newspaper Association of America, "Trends and Numbers: Circulation."
33. *Presstime*, "Newspaper Jobs Now Total 432,000; Most in Nation," 50
34. Ibid.
35. Emery and Emery, *The Press and America: An Interpretive History of the Mass Media*, 4th ed., 119–123.
36. Neuharth, "Opportunities for Newspapers Will Abound in Exciting '80s," 2. Allen H. Neuharth was chairman and president of Gannett Inc.
37. Ibid.
38. Toffler, *The Third Wave* and *Previews and Premises*.
39. Naisbitt, *Megatrends: Ten New Directions Transforming Our Lives*.
40. Watts, "The '80s: Telecommunications," 41. Douglas Watts was a member of the American Newspaper Publishers Association's legal staff.
41. Maney, *Megamedia Shakeout: The Inside Story of the Leaders and Losers in the Exploding Communications Industry*, 167.
42. Newspaper Association of America, "Trends and Numbers: Advertising Expenditures."

43. Maxwell and Wanta, "Advertising Agencies Reduce Reliance on Newspaper Ads," par. 8

44. Newspaper Association of America, "Trends and Numbers: Advertising Expenditures."

45. U.S. Census Bureau, *Statistical Abstract of the United States: 2008*, 704.

46. Newspaper Association of America, "Trends and Numbers: Circulation."

47. U.S. Census Bureau, *Statistical Abstract of the United States: 2011*, 714.

48. Ibid., 709.

49. U.S. Bureau of Labor Statistics, "Employment Projections 2008–2018."

50. Ibid.

51. U.S. Bureau of Labor of Statistics, "Occupational Employment Statistics."

52. Walker, "Another Role for Turner," D2.

53. Bogart, "Expect No Substitutes: Printed Newspapers Will Ride Out the Electronic Wave," 41. Statements were attributed to Roger Fidler, a pioneer in Knight-Ridder's electronic endeavors, and Frank M. Daniels III, the former executive editor of *News & Observer* in Raleigh, NC.

54. Crichton, "Mediasaurus," par 1.

55. *Presstime*, "Peters: Get Crazy, or Else," 30. Tom Peters is best known as the author of *In Search of Excellence: Lessons from America's Best-Run Companies.*

56. Peskin, "Slaying 'The Mediasaurus:' An Editor Debunks Michael Crichton's View of Media Obsolescence," 52. Dale Peskin was an assistant managing editor at the *Detroit News* and a former editor of the magazine for the Society of Newspaper Design.

57. Compaine, "The '80s: An Overview," 6.

58. Ibid., 7–8.

59. Potter, "The Deja View: Publishers' Roles in the Development of, and Adaptation to, Broadcast Offer Uplift in the New-Media Age," 28. Observations and statements attributed to Scott Whiteside, an executive for Times Mirror Co. and later at Cox Enterprises Inc.

60. Ibid., 29.

61. Bovee, *Discovering Journalism*, 51.

62. Udell, *The Economics of the American Newspaper*, 14.

63. Buffett, "Chairman's Letter—1984," http://www.berkshirehathaway.com/letters/1984.html. Warren Buffett was chairman of Berkshire Hathaway Inc., which was the owner of the *Buffalo Evening News.*

64. Morton, "When Newspapers Eat Their Seed Corn," 52. John Morton was a former newspaper reporter who became a financial analyst specializing in the newspaper and other media-related industries.

65. Overholser, "Peril, Paranoia and Promise," 54.

66. Patterson, "Buffett Sees 'Unending Losses' for Many Newspapers," par. 2. Statement attributed to Warren Buffett, chairman of Berkshire Hathaway Inc.

67. Holm and Stewart, "In Deal, Buffett Departs from Type," C1.

68. Mayer, "The Polls—Poll Trends: The Rise of the New Media," 124.
69. Ibid.
70. Rice, "Artifacts and Paradoxes in New Media," 26.
71. Manovich, *The Language of New Media*, 47.
72. Koziol, "Remix the Media Mix," par. 1.
73. Green and Haddon, *Mobile Communications: An Introduction to New Media*, 1.
74. Kiousis, "Interactivity: A Concept Explication," 371.
75. Ibid., 357.
76. Downes and McMillan, "Defining Interactivity: A Qualitative Identification of Key Dimension," 157.
77. Ibid., 157–158.
78. Ibid., 173.
79. Stromer-Galley, "Interactivity-as-Product and Interactivity-as-Process," 393.
80. Ibid.
81. Bucy, "Interactivity in Society: Locating an Elusive Concept," 376.
82. Hartley, *Communication, Cultural and Media Studies: The Key Concepts*, 5.
83. Baldwin, McVoy, and Steinfield, *Convergence: Integrating Media, Information and Communication*, 1.
84. Dominick, *The Dynamics of Mass Communication*, 19.
85. Ibid., 20.
86. Ibid., 21.
87. Ibid.

1. Videotext and the Birth of Online Newspapers

1. Gillen, "Letter from the President: The Waiting Is Over; Touch the Future," 2.
2. Kyrish, "Lessons from a 'Predictive History': What Videotex Told Us about the World Wide Web," 10.
3. This book uses the spelling "videotext" rather than "videotex" (unless quoting an author who used the alternate spelling). The review of sources found videotext with a "t" at the end to be the common spelling in many contemporary articles. However, videotex was the official spelling used by the Videotex Industry Association (VIA) and the International Telegraph and Telephone Consultative Committee (CCITT), an advisory body of the International Telecommunications Union (ITU). Therefore, many of the sources cited used the spelling without the final "t" as this was the preferred spelling when videotext was a deployed technology in the 1980s and early 1990s.
4. Gillen, "Letter from the President," 2.
5. Kyrish, "Lessons from a 'Predictive History,'" 12.
6. Ibid., 11.
7. Ibid.

8. Ibid.
9. Kyrish, "From Videotext to the Internet," 10.
10. Rambo, "New Services Stir Variety of Questions on Marketing," 24.
11. Ibid., 26.
12. Butler and Kent, "Potential Impact of Videotext on Newspapers," 7.
13. Udell, *The Economics of the American Newspaper*, 144.
14. Kyrish, "Lessons from a 'Predictive History,'" 25–26.
15. Mowshowitz, "Scholarship and Policy Making: The Case of Computer Communications," 4.
16. Russell, "Advertising in an Electronic World," 52.
17. Rosenblatt, "Machine of the Year: A New World Dawns," 13.
18. Bagdikian, *The New Media Monopoly*, 57.
19. Flichy, "The Construction of New Digital Media," 33.
20. Boczkowski, "The Mutual Shaping of Technology and Society in Videotex Newspapers," 258.
21. Ibid.
22. *Viewtron Magazine*, "Viewtron's Roots Traced; From Telephone to TV to Videotex," 22.
23. Sigel, *The Future of Videotext: Worldwide Prospects for Home/Office Electronic Information Services*, 2.
24. Ibid., 1. See footnote 3 for an explanation regarding spelling.
25. Ibid.
26. Ibid.
27. Neustadt, *The Birth of Electronic Publishing: Legal and Economic Issues in Telephone, Cable and Over-the-Air Teletext and Videotext*, 1.
28. Aumente, *New Electronic Pathways: Videotex, Teletext, and Online Databases*, 14.
29. Ibid., 17.
30. Ibid.
31. Sigel, *The Future of Videotext*, 3.
32. Neustadt, *The Birth of Electronic Publishing*, 1.
33. Ibid.
34. Kist, *Electronic Publishing: Looking for a Blueprint*, 11.
35. Ibid.
36. Lerner, "Electronic Publishing," 111.
37. Cuadra, "A Brief Introduction to Electronic Publishing," 29–34.
38. Gurnsey, *Electronic Document Delivery—III: Electronic Publishing Trends in the United States and Europe*.
39. Udell, *The Economics of the American Newspaper*, 94.
40. Ibid.
41. Compaine, *The Newspaper Industry in the 1980s: An Assessment of Economics and Technology*, 208.

42. Ibid.
43. Picard, "Changing Business Models of Online Content Services," 63.
44. Kist, *Electronic Publishing*, 14.
45. Ibid.
46. Standera, *The Electronic Era of Publishing: An Overview of Concepts, Technologies and Methods*, 1.
47. *Editor & Publisher*, "It's a Gee Whiz World," 8.
48. Boczkowski, "The Mutual Shaping of Technology and Society in Videotex Newspapers," 258–259.
49. Smith, *Goodbye, Gutenberg: The Newspaper Revolution of the 1980s*, 3.
50. Weingarten, "Testimony before Congress: New Information Technology and Copyrights," 4. Weingarten was manager of the Communication and Information Technologies Program within the federal government's Office of Technology Assessment.
51. Alber, *Videotex/Teletext: Principles and Practices*, xi.
52. Beniger, *The Control Revolution: Technological and Economic Origins of the Information Society*, 21.
53. Leavitt and Whisler, "Management in the 1980s," 41–48.
54. Brynjolfsson and Saunders, *Wired for Innovation: How Information Technology Is Reshaping the Economy*, 53.
55. Ibid., 62.
56. Toffler, *Future Shock*.
57. Naisbitt, *Megatrends: Ten New Directions Transforming Our Lives*.
58. Bagdikian, *The Information Machines: Their Impact on Men and the Media*, 184.
59. Ibid.
60. Ibid., 205.
61. Ibid., 184.
62. Ibid., 196.
63. Ibid., 197.
64. Mowshowitz, "Scholarship and Policy Making," 3.
65. Ibid.
66. Ibid., 4.
67. Arnold and Arnold, "Vectors of Change: Electronic Information from 1977 to 2007," par. 35.
68. Ibid., par. 16.
69. Ibid., par. 29.
70. Standera, *The Electronic Era of Publishing*, 48.
71. Mowshowitz, "Scholarship and Policy Making," 3.
72. Ibid., 2.
73. Laakaniemi, "The Computer Connection: America's First Computer-Delivered Newspaper," 62.

74. Ibid., 63.

75. Kyrish, "From Videotext to the Internet," 12.

76. Ibid., 16.

77. Shedden, "New Media Timeline (1980)."

78. Ibid.

79. Shedden, "New Media Timeline (1985)."

80. Wright, "Electronic Publishing: How to Use It and Why," 25.

81. Shedden, "New Media Timeline (1983)."

82. Chanin, "Officials of England's 'Electronic Newspaper' Predict No Early Demise of Print Media," 6.

83. Genovese, "Electronics Boom a Gamble or Sure Bet?," 5.

84. Ibid.

85. Alber, *Videotex/Teletext*, 4.

86. Lancaster, *Toward Paperless Information Systems*, xi.

87. Ibid., 1.

88. Ibid., 161.

89. Compaine, *The Newspaper Industry in the 1980s*, 209.

90. Boczkowski, "The Mutual Shaping of Technology and Society in Videotex Newspapers," 258–259.

91. Carlson, "The News Media's 30-Year Hibernation," 69.

92. Russell, "Advertising in an Electronic World," 52.

93. Ibid.

94. Compaine, *The Newspaper Industry in the 1980s*, 213.

95. Kauffman, "Newspapers See Opportunities in Electronic Classified Ads," 35. Kauffman was president of the Newspaper Advertising Bureau.

96. Ibid.

97. *Folio: The Magazine for Magazine Management*, "Advertising Agencies Link Ads, Editorial in Electronic Publications," par. 12.

98. Ibid., pars. 14–16. The statement was attributed to Timothy Conner, director of new electronic media at Young & Rubicam.

99. Ibid., par. 7.

100. Boczkowski, "The Mutual Shaping of Technology and Society in Videotex Newspapers," 258.

101. Udell, *The Economics of the American Newspaper*, 38.

102. Boczkowski, "The Mutual Shaping of Technology and Society in Videotex Newspapers," 258.

103. Ibid.

104. Ibid.

105. Ibid.

106. Genovese and Rambo, "'BISON' Bites Dust, but 'STAR-Text' Is Launched," 16.

107. Laakaniemi, "The Computer Connection," 61.

108. Ibid., 62.

109. Ibid.

110. Ibid.

111. Rittenhouse, "The Market for Wired City Services," 2.

112. Laakaniemi, "The Computer Connection," 64.

113. Ibid., 65.

114. Ibid., 67.

115. Ibid., 63–64.

116. Shedden, "New Media Timeline (1980)."

117. Hilder, "Will Public Like Newspapers on TV?," 16B

118. Hecht, "Information Services Search for Identity," 61.

119. *Presstime*, "CompuServe Evaluation Marks End to Second Year of Operation," 17. Statement attributed to Robert J. Bettencourt, business development manager for the *Virginian-Pilot* and the *Ledger-Star*.

120. Wellborn, "A World of Communications Wonders," 59. Statement attributed to Charles Lecht, chairman of Lecht Sciences Inc., a communications consulting firm in New York.

121. Shedden, "New Media Timeline (1982)."

122. Barker, "Back to the Future," 45.

123. Ibid.

124. Ibid.

125. Friedman, "Videotex Languor Claims Victims," 56.

126. Genovese, "Newspapers Keep Marching to Videotex," 28.

127. Barker, "Back to the Future," 45.

128. Ibid.

129. Mencke, "Farewell to a Good Friend," pars. 1–2.

130. Boczkowski, *Digitizing the News: Innovation in Online Newspapers*, 25.

131. Ibid.

132. Ibid.

133. *Viewtron Magazine*, "Viewtron's Roots Traced; From Telephone to TV to Videotex," 22.

134. Genovese, "All Eyes Are on Viewtron Screen Test," 22.

135. Finberg, "Before the Web, There Was Viewtron," par. 13. Statement attributed to Norman Morrison, vice president of the Knight-Ridder subsidiary Viewdata Corp.

136. Sigel, *The Future of Videotext*, 67–68.

137. Ibid., 68–69.

138. *Viewtron Magazine*, "Merchants by Category," 7.

139. *Viewtron Magazine*, "The 'Electronic Mall' Arrives; All Under One Roof (Yours)," 6.

140. Burgess, "Firms Face Questions of Technology's Timing, Cost," par. 5.

141. Ghosh, "Videotex Systems," 21.

142. Ibid.

143. Genovese, "Viewtron Partners Disappointed over Lackluster Sales," 40. Statement attributed to Robert H. Phelps, vice president of Affiliated Publications

144. Ibid.

145. Ghosh, "Videotex Systems," 21.

146. Millman, "Videotext Age Comes to an End: Viewtron Folds; Knight-Ridder Pulls Plug on Unit," par. 14.

147. Ghosh, "Videotex Systems," 21.

148. Pryor, "The Videotex Debacle," 41.

149. Wright, "Electronic Publishing: How to Use It and Why," 25.

150. Ibid.

151. Ibid.

152. Ibid.

153. Pryor, "The Videotex Debacle," 41.

154. Ibid., 42.

155. Ibid.

156. Kyrish, "From Videotext to the Internet," 25.

157. Pryor, "The Videotex Debacle," 42. Larry Pryor had been the managing editor of the Times Mirror Gateway system from 1982 to 1986.

158. Ibid.

159. Friedman, "Videotex Languor Claims Victims," 56. Statement attributed to Phillip J. Meek, publisher of the *Fort Worth Star-Telegram*.

160. *Presstime*, "Telecommunications Notes," 17.

161. Kyrish, "From Videotext to the Internet," 25.

162. Genovese and Rambo, "'BISON' Bites Dust, but 'STAR-Text' Is Launched," 16.

163. *Presstime*, "CompuServe Evaluation Marks End to Second Year of Operation," 17.

164. Henke and Donahue, "Teletext Viewing Habits and Preferences," 545.

165. Ibid.

166. Stevenson, "Videotex Players Seek a Workable Formula," par. 12.

167. Ibid., par. 3.

168. Ibid., par. 7.

169. Raymond, "Why Videotex Is (Still) a Failure," 35.

170. Ibid.

171. Ibid., 36.

172. Ibid.

173. Atwater, Heeter, and Brown, "Foreshadowing the Electronic Publishing Age: First Exposures to Viewtron," 813.

174. Ibid., 814.

175. *U.S. News & World Report*, "On Horizon: Home Computers with a Gift of Gab (Interview with Howard Anderson)," 58.

176. Finberg, "Before the Web, There Was Viewtron," pars. 16–20.

177. Ibid.

178. Finberg, "Viewtron Remembered Roundtable," par. 12. Statement was attributed to Phil Meyer, journalism professor at the University of North Carolina and former Viewtron director of market research.
179. Raymond, "Why Videotex Is (Still) a Failure," 36.
180. Finberg, "Viewtron Remembered Roundtable," par. 34. Statement attributed to Meyer.
181. Ibid., par. 6. Statement attributed to Rita Haugh Oates, Viewtron's education editor.
182. Ibid., par. 21. Statement attributed to Reid Ashe.
183. Ibid., par. 24. Statement attributed to Bob Cochnar, Viewtron's managing editor.
184. Ibid., par. 13. Statement attributed to Reid Ashe.
185. Boczkowski, "The Mutual Shaping of Technology and Society in Videotex Newspapers," 263.
186. Kyrish, "Lessons from a 'Predictive History,'" 26.
187. Boczkowski, "The Mutual Shaping of Technology and Society in Videotex Newspapers," 263.

2. The Newspaper Industry's Brief Cable Television Strategy

1. Effros, "The Reality Behind the Cable TV Boom," 25. Stephen Effros was executive director of the Community Antenna Television Association.
2. Whitney, "All Signs Point to Home-Computer Revolution," 47.
3. *Computer Industry Almanac*, "25-Year PC Anniversary Statistics," 14.
4. *Computer Industry Almanac*, "The U.S. Now Has One Computer for Every Three People," 28.
5. Genovese, "PCs No Longer Seen as a Threat," 20.
6. Ibid.
7. Ibid.
8. Boczkowski, *Digitizing the News: Innovation in Online Newspapers,* 33.
9. Besen and Crandall, "The Deregulation of Cable Television," 81.
10. U.S. Census Bureau, *Statistical Abstract of the United States: 1980,* 589.
11. Ibid.
12. Genovese, "Newspapers Channel Interest in Cable TV," 4.
13. Besen and Crandall, "The Deregulation of Cable Television," 79.
14. Moss, "Cable Television: A Technology for Citizens," 701.
15. Besen and Crandall, "The Deregulation of Cable Television," 78.
16. Genovese, "Newspapers Channel Interest in Cable TV," 4.
17. Besen and Crandall, "The Deregulation of Cable Television," 78.
18. Ibid., 110.
19. Noam, "Towards an Integrated Communications Market: Overcoming the Local Monopoly of Cable Television," 235.
20. Ibid., 236.

21. Genovese, "Newspapers Channel Interest in Cable TV," 4.

22. Besen and Crandall, "The Deregulation of Cable Television," 110.

23. Rosenfeld, "Videotext vs. Newspapers: The Press's $64,000 Ifs," 28. Arnold Rosenfeld was the executive editor of the *Dayton Daily News* and chairman of the American Society of Newspaper Editors' Technology Committee.

24. *Newspaper Controller*, "A Look at Electronic Publishing," 7. Statement attributed to Larry Pfister, who was Time Inc.'s vice president of Video Information Services.

25. Ibid.

26. Genovese, "Newspapers Channel Interest in Cable TV," 5.

27. These figures were compiled from four annual capital expenditure surveys conducted by the ANPA Research Institute and published in *Presstime* (July 1982, 49; July 1983, 57; May 1984, 63; May 1985, 66. For information on ANPA, see note 32 in the Introduction).

28. Genovese, "Newspapers Channel Interest in Cable TV," 7.

29. Ibid.

30. Rambo, "Clouding the Future? Electronic Publishers Face Legal Questions on Variety of Issues," 5. Statement attributed to Douglas Watts, the ANPA Legislative Counsel.

31. Ibid.

32. Ibid. Statement attributed to Thomas McKnight, a communications industry consultant who was a former executive with Gannett Inc.

33. Genovese, "Newspapers Channel Interest in Cable TV," 5. Statement attributed to Gerald Moriarity, who was publisher the *Globe-Gazette* in Mason City, Iowa, and earlier publisher of the *Ottumwa Courier*.

34. Criner and Gallagher, "Newspaper-Cable TV Services: Current Activities in Channel Leasing and Other Local Service Ventures," A3.

35. Ibid.

36. *Presstime*, "More Papers Entered Cable TV for Competition than Profits," 30.

37. Ibid.

38. Becker, Dunwoody, and Rafaeli, "Cable's Impact on Use of Other News Media," 127.

39. Ibid., 130.

40. McCombs, "Mass Media in the Marketplace," 1–104.

41. McCombs and Eyal, "Spending on Mass Media," 153–158.

42. Weiss, "Effects of the Mass Media of Communication," 77–195.

43. Robinson, "Television and Leisure Time: A New Scenario," 120–130.

44. Becker et al., 131.

45. Genovese, "Newspapers Channel Interest in Cable TV," 7. Statement attributed to Charles Miller, publisher of *The Recorder* in Amsterdam, NY.

46. *Presstime*, "Electronic Publishers Describe How They Do It," 18. Statement attributed to Lee Porter, publisher of the *Shawnee News-Star* in Oklahoma.

47. Ibid.
48. Ibid.
49. *Newspaper Controller*, "A Look at Electronic Publishing," 7.
50. Reed, "Cabletext: Is It Newspapering?" 23. David Reed was director of the Lexington, Kentucky, *Herald-Leader's* TelePress unit.
51. Ibid.
52. Ibid.
53. Wicklein, "The Scary Potentials for the Overcontrol of Information," 13. John Wicklein had been a reporter and editor at the *New York Times* and dean of the School of Public Communication at Boston University.
54. Ibid., 15.
55. Ibid., 14.
56. Ibid., 15.
57. Rambo, "It's Still a 'Maybe' Market for New Technologies," 20.
58. Genovese, "Latest Tack for Papers in Cable TV Is to Video," 20.
59. Rambo, "It's Still a 'Maybe' Market for New Technologies," 22.
60. Ibid.
61. Ibid., 21. Statement attributed to Mark Atkinson, the general manager of cable operations at the *Leader-Telegram* in Eau Claire, Wisconsin.
62. Ibid., 22.
63. Potter, "Early Newspaper Efforts as Information Providers on Cable Were Enthusiastic and Imaginative but Fell Short," 11.
64. Ibid., 10. Statement attributed to Robert M. White II, who had been publisher of the *Mexico Ledger* in Missouri.
65. Potter, "A Local Look for Cable-TV News," 9. Statement attributed to Robert J. Gremillion, vice president and general manager of a Tribune Co. subsidary developing a local cable news project in Chicago.
66. Ibid.
67. *Presstime*, "Cable TV Start-Ups Show Decline Since Peak in 1982," 24. Statement attributed to Kathleen Criner, ANPA's director of telecommunications affairs.
68. *Newspaper Controller*, "The Economics of Telecommunications," 16. Statement attributed to Kathleen Criner.
69. Effros, "The Reality Behind the Cable TV Boom," 25.
70. Ibid.
71. Rambo, "Newspaper Companies Go Back to Basics," 22.
72. *Newspaper Controller*, "The Economics of Telecommunications," 16. Statement attributed to Kathleen Criner.
73. Ibid.
74. Ibid., 15.
75. *Presstime*, "Electronic Publishers Describe How They Do It," 18.

76. Rambo, "Newspaper Companies Go Back to Basics," 28. Statement attributed to Dean Singleton, president of Media News Group, which then owned 26 U.S. daily newspapers.

77. Ibid., 27. Statement attributed to Erwin Potts, executive vice president of McClatchy Newspapers.

78. *Presstime*, "'No Revenue Potential' Ends Cable Service," 12. Statement attributed to James L. Whyte, president and general manager of the newspaper's parent, Florida Publishing Co.

79. Ibid.

80. Friedman, "Cable TV: Now Seems to Be a Good Time to Sell or Consolidate System, and Newspapers Are Reaping Benefits Both Ways," 5.

81. Ibid., 6.

82. Ibid., 5. Statement attributed to Frank Washington, McClatchy's vice president for electronic communications.

83. Rambo, "Newspaper Companies Go Back to Basics," 22. Statement attributed to Erwin Potts, McClatchy's executive vice president.

84. Ibid. Statement attributed to Ellen Gibbs, president of Communications Resources Inc.

85. Ibid.

86. Ibid., 22.

87. Ibid., 29.

88. Friedman, "Cable TV: Now Seems to Be a Good Time To Sell or Consolidate Systems, and Newspapers Are Reaping Benefits Both Ways," 6. Statement attributed to Joseph Hays, vice president of corporate communications.

89. Rambo, "Newspaper Companies Go Back to Basics," 27.

90. Ibid.

91. Potter, "Early Newspaper Efforts as Information Providers on Cable Were Enthusiastic and Imaginative but Fell Short," 11. Statement attributed to Timothy Brennan, vice president of electronic information products for Tribune Media Services Inc.

92. Rambo, "Newspaper Companies Go Back to Basics," 27. Statement attributed to Erwin Potts, McClatchy's executive vice president.

93. Ibid.

94. Patten, *Newspapers and New Media,* 43.

95. Ibid.

96. Finnegan and Viswanath, "Community Ties and Use of Cable TV and Newspapers in a Midwest Suburb," 458.

97. Rambo, "Newspaper Companies Go Back to Basics," 27. Statement attributed to Erwin Potts, McClatchy's executive vice president.

98. Friedman, "Cable TV: Now Seems to Be a Good Time to Sell or Consolidate Systems, and Newspapers Are Reaping Benefits Both Ways," 6. Statement

attributed to Frank Washington, McClatchy's vice president of electronic communications.

99. Rambo, "Newspaper Companies Go Back to Basics," 23. Statement attributed to Christopher Shaw, chairman of Henry Ansbacher Inc.

100. Ibid. Statement attributed to Bruce Thorp, newspaper securities analyst with Lynch, Jones & Ryan.

101. Ibid., 30. Statement attributed to James Longson, vice president of corporate development with Tribune Co.

102. Ibid. Statement attributed to Lee Dirks of Lee Dirks & Associates.

103. Gomery, "Media Economics: Terms of Analysis," 58.

3. Newspapers React to Fear of Telecommunication Dominance

1. American Telephone and Telegraph Company, "Directory and Public Services," 12.

2. Ibid.

3. Criner and Wilson, "Telecommunications History Is Short but Stormy," 24.

4. Ibid.

5. Noam, "Towards an Integrated Communications Market: Overcoming the Local Monopoly of Cable Television," 245.

6. Ibid., 257.

7. Noam's comments also foreshadow the events of the early 1990s, when the emergence of the Internet as a consumer medium was also unplanned and challenged the regulatory environment in much the same way.

8. Smith, *Goodbye, Gutenberg: The Newspaper Revolution of the 1980s*, 306.

9. Ibid.

10. Rambo, "AT&T's Home Video Plans Eyed Warily," 48.

11. Brown, "Report of the Chairman of the Board," 2.

12. Rambo, "AT&T's Home Video Plans Eyed Warily," 48.

13. Ibid., 49.

14. Ibid., 48.

15. Maguire, "How the Diversity Principle Became the Law of the Land," 23. W. Terry Maguire was the ANPA's vice president and general counsel.

16. Ibid.

17. Ibid.

18. Rambo, "AT&T's Home Video Plans Eyed Warily," 48. Statement attributed to Robert Marbut, CEO of Harte-Hanks Communications Inc., who was chairman of the industry's telecommunications committee.

19. Fink, *Strategic Newspaper Management*, 16.

20. Ibid.

21. Rambo, "Legislative Activity Focusing on AT&T Deregulation Bills," 44. Statement attributed to Jerry Friedham, who was executive vice president and general manager of the ANPA.

22. According to the NAA, "the Newspaper Preservation Act (NPA) permits newspapers in the same city to enter joint operating arrangements (JOAs)—combining business operations—when one of the participants is in probable danger of financial failure. The Act has sustained competition by enabling JOA newspapers to maintain the ability to publish separate and editorially independent publications by combining their business operations (printing, distribution, and advertising and subscription sales) into a single enterprise." http://www.naa.org/Public-Policy/Legal-Affairs/Business-Operations.aspx.

23. LeGates, "Many Noses Being Poked Under Information Tent," 25. John LeGates was director of Harvard University's Program on Information Resources Policy.

24. Ibid., 23.

25. Ibid., 25.

26. Ibid.

27. Ibid., 24.

28. Maguire, "How the Diversity Principle Became the Law of the Land," 24.

29. Ibid.

30. Marbut, "Confronting the Telecommunications Issue," 29.

31. Ibid., 31.

32. Kinsley, "The Latest Moos: Newspaper Publishers Protect Their Cash Cows," 21.

33. Ibid., 22.

34. Ibid.

35. Cole, Barry G., *After the Breakup: Assessing the New Post-AT&T Divestiture Era,* 8–11.

36. Rambo, "Court Approves AT&T Consent Decree," 10.

37. Cole, Barry G., *After the Breakup,* 11.

38. Rambo, "Court Approves AT&T Consent Decree," 10.

39. Ibid., 11.

40. Maguire, "How the Diversity Principle Became the Law of the Land," 23.

41. Rambo, "Court Approves AT&T Consent Decree," 10.

42. *Presstime,* "Convention Report: AT&T's Brown Rebuts Press Concerns," 21.

43. Ibid.

44. Ibid., 21–22.

45. Rambo, "Court Approves AT&T Consent Decree," 10.

46. Mowshowitz, "Scholarship and Policy Making: The Case of Computer Communications," 5.

47. Rambo, "Divestiture: How Will Breakup of Ma Bell Affect Newspapers?" 10–11.

48. Ibid.

49. Ibid., 10.
50. Ibid. Statement attributed to Christopher J. Burns, who was identified as an independent consultant.
51. Criner and Wilson, "An Uncertain Marketplace Takes Telecommunications Toll," 23.
52. Ibid.
53. Criner, "Newspapers at Mid-Decade and Beyond: Telecommunications," 26.
54. Ibid.
55. Fink, *Strategic Newspaper Management*, 16.
56. Rambo, "Audiotex: Dial-Up Services Now Provide Weather, Sports Scores; Could Bridge the Gap to More Complex Systems," 28.
57. Ibid.
58. Genovese, "'VIS'ion of the Future: 'Voice Information' Offers a Relatively Low-Cost Way to Dabble in New Telecommunications Services," 24. The observation was attributed to Robert Johnson, publisher of *Newsday* and chairman of the ANPA telecommunications committee.
59. Potter, "How to Build Your Own Audiotex System," 10.
60. Friedman, "Waivers: Bell Regionals Get Green Light to Expand into New Areas," 14.
61. Ibid.
62. *Presstime*, BellSouth, ANPA Clash Over Progress of Gateway," 64.
63. *Presstime*, "'Diversity Principle' Reaffirmed," 42.
64. Johnson, "Use It or Lose It: Electronic Publishing; Between Now and 1990, Newspapers Must Act to Convince the Court They Are Serious About Offering High-Tech Information Services," 10. Robert Johnson was publisher of *Newsday* and chairman of the ANPA telecommunications committee.
65. Ibid., 12.
66. Ibid., 10.
67. *Presstime*, "Telecommunications Technology Beckons, But Is It a Boon or a Bane?" 46. Frank Blethen was president and chief executive officer of the *Seattle Times* and succeeded Johnson as chairman of the trade association's telecommunications committee.
68. *Presstime*, "Voice-Information Services Provide Successful Newspaper Ventures," 43.
69. *Presstime*, "Three BOCs Rebuffed in Bid to Offer Electronic Yellow Pages," 61.
70. *Presstime*, "AT&T Petitions Court for Removal of Electronic Publishing Curbs," 70.
71. Aumente, *New Electronic Pathways: Videotex, Teletext, and Online Databases*, 11.
72. Veronis, "The Bells Are Ringing: Newspapers and Telephone Companies Remain Adversaries over Information Services, and This Year May Be Critical," 14.
73. Ibid., 15.

74. *Presstime*, "Regional Bell Companies Mount Strong Fight to Enter Electronic Publishing," 42.

75. Kahaner, "'Baby' Bells Grow into Giants: Newspapers Fear RBOC's May Become Information Providers over Their Own Lines," 18. Larry Kahaner was a founding editor of *Communications Daily* newsletter.

76. Potter, "Changed Marketplace, Changed Attitudes: Newspapers and the Bells Have Changed Their Tone as They Seek Common Ground," 28–29.

77. *Presstime*, "Telecommunications Is a 'Pathway to the Future,'" 35.

78. Potter, "Papers Battle RBOC's on All Fronts," 54.

79. Ibid.

80. Ibid., 56.

81. Ibid.

82. Potter, "Bells' Worries Range Far and Wide," 28. Statement attributed to Laird Walker, vice president of federal relations for US West.

83. *Presstime*, "Voice-Information Services Provide Successful Newspaper Ventures," 43

84. Ibid.

85. Light, "Facsimile: A Forgotten 'New Medium' from the 20th Century," 356.

86. Ibid., 358.

87. Scully, "Media Market: The Bells Toil for Them," par. 4. Statement attributed to Nicholas Penniman, publisher of the *St. Louis Post-Dispatch*.

88. Ibid. Statement attributed to Dick Holzapfel, assistant vice president for information services at BellSouth.

89. Potter, "Changed Marketplace, Changed Attitudes: Newspapers and the Bells Have Changed Their Tone as They Seek Common Ground," 29.

90. Crowe and Jones, "Countdown to N11," 33.

91. Potter, "From Concept to Strategic Plan: Dow Jones Makes Some Connections," 6.

92. Potter, "Changed Marketplace, Changed Attitudes," 29.

93. Ibid., 27–28. Statement attributed to Thomas Pace, alliances manager at Dow Jones & Co.

94. Ibid., 30. Statement attributed to Laird Walker, vice president of federal relations for US West Inc.

95. *Presstime*, "Assessing the Challengers," 36.

96. Erburu, "We Can Win Information Services Fight Because Newspaper Position Is Right," 5.

97. Potter, "Broader Fight Looms in House Over Restrictions on RBOCs," 38.

98. Bogart, "There's Plenty To Do Before the Unthinkable Happens," 58.

99. *Presstime*, "Newspapers Must Adopt New Technology to Improve Role of 'Print Gateway,'" 30.

100. *Presstime*, "Papers Must Deliver Information in Many New Ways, Erburu Says: Traditional Print Publication Is Not for Everyone," 26. Robert Erburu was

chairman and chief executive officer of the Times Mirror Co. and chairman of ANPA.

101. America, "RBOCs Battle Enters New Phase," 45.

102. *Presstime*, "Tomorrow's Technology: Be There Now," 40. Statement attributed to Scott Whiteside, vice president of strategic planning and advanced systems at the *Baltimore Sun*, who later held strategic positions at Times Mirror and Cox Enterprises.

4. Newspapers Embrace Proprietary Online Services

1. Cameron, et al., "Electronic Newspapers: Toward a Research Agenda," 3.

2. Ibid., 34.

3. Burgess, "Firms Face Questions of Technology's Timing, Cost," par. 37.

4. Ibid., par. 6.

5. Rheingold, *The Virtual Community: Homesteading on the Electronic Frontier*, xx.

6. Ibid., xv–xvi.

7. Booker, "Consumer Videotex: The Perilous Path (Prospects for the Videotex Industry in the U.S.)," pars. 1, 7.

8. Ibid., par. 5.

9. Kinsley, "America Is Losing the Information Revolution," pars. 1–3.

10. Booker, "Consumer Videotex," par. 20. Statement attributed to C. Sidney Boren, a vice president with BellSouth.

11. Conhaim, "Developing Videotex as a Consumer Medium," par. 3.

12. The phrase was ubiquitous in the late 1990s; it was also used as the title of a popular film released in 1998 starring Meg Ryan and Tom Hanks.

13. Reid, *Architects of the Web: 1,000 Days That Built the Future of Business*, 14.

14. Kyrish, "From Videotext to the Internet: Lessons from Online Services 1981–1996," 15.

15. Ibid.

16. McKenna, "The Future Is Now: Newspapers Are Overcoming Their Fears of Technology and Launching a Wide Array of Electronic Products," par. 44.

17. Carter, "Online Advertising," 38–39. Statement attributed to Frederick J. Tuccillo, director of electronic publishing for *Newsday* in New York.

18. Underwood, "Reinventing the Media: The Newspapers' Identity Crisis," 25. Doug Underwood was a communications professor at the University of Washington and a former newspaper journalist.

19. Ibid., 24–25.

20. Weintrob, "An Eerily Prescient 1994 Vision of the 2010 iPad," par. 1.

21. Underwood, "Reinventing the Media," 25.

22. Glaberson, "The Media Business: In San Jose, Knight-Ridder Tests a Newspaper Frontier," par. 10.

23. Markoff, "17 Companies in Electronic News Venture," par. 12. Statement attributed to Robert Ingle, executive editor of the *San Jose Mercury News*, part of the Knight-Ridder chain that created Viewtron.

24. McKenna, "The Future Is Now," par. 32. Statement attributed to Gerry Barker, online service marketing director at the *Fort Worth Star-Telegram*.

25. Underwood, "Reinventing the Media," par. 16. Statement attributed to Gerry Barker.

26. McKenna, "The Future Is Now," par. 12. Statement attributed to Frank Hawkins, a vice president at Knight-Ridder.

27. Wilson and Igawa, "On Our Minds: Worry, Worry, Worry, or Innovate?" 27.

28. Ibid.

29. Markoff, "17 Companies in Electronic News Venture," pars. 1–4.

30. Massachusetts Institute of Technology, "Consortium to Probe News Presentation," par. 5. Statement attributed to Nicholas Negroponte, director of MIT's Media Laboratory.

31. McKenna, "Where the Action Is," par. 2. Statement attributed to Jack Driscoll, a former editor of the *Boston Globe*, who was managing the newspaper's electronic information initiatives.

32. Boczkowski, *Digitizing the News: Innovation in Online Newspapers*, 36.

33. Ibid. The statement was attributed to Frank Hawkins, a vice president at Knight-Ridder.

34. Morton, "Papers Will Survive Newest Technology," 48.

35. Potter, "Media Merger Mania II: The Quest for Content," 29. Statement attributed to Gordon Mendica, vice president of operations and planning, at the New York Times Co.

36. Ibid. Statement attributed to Frank Hawkins, a vice president with Knight-Ridder Inc.

37. Ibid. Statement attributed to Eugene Quinn, general manager of Chicago Online, the electronic operations of the *Chicago Tribune*.

38. Boczkowski, *Digitizing the News*, 49.

39. *Presstime*, "Print Still Prime," 34. Statement attributed to EuGene L. Falk, executive vice president and general manager of the *Los Angeles Times*.

40. Ibid., 39.

41. Potter, "A Guide to Online Avenues," S2. Statement attributed to Donald K. Brazeal, editor and publisher of the Washington Post Co.'s online media subsidiary.

42. *Presstime*, "Panel's Advice: Get in Gear; Content, Identity Drive Development," 27. Statement attributed to Charles T. Brumback, chairman, president and CEO of Tribune Co.

43. Toner, "Getting on Boards," 47–48. Statement attributed to David Carlson, director of the new media lab at the University of Florida's College of Journalism and Communications. He had earlier worked for the *Albuquerque Tribune*'s electronic edition.

44. Moeller, "The High-Tech Trib," par. 59. Statement attributed to Jim Longson, corporate vice president of technology at Tribune Co.

45. McKenna, "The Future Is Now," par. 16. Statement attributed to David Easterly, president of Cox Newspapers.

46. *Presstime*, "Print Still Prime," 34. Statement attributed to Marty Petty, vice president of sales and marketing for the *Hartford Courant*.

47. *Presstime*, "Partner Power and the Next Media Age," 38. Statement attributed to Susan Decker, a newspaper industry analyst with the brokerage firm Donaldson, Lufkin & Jenrette.

48. Ibid., 39. Statement attributed to Susan Bokern, associate director of new business for Gannett's *USA Today*.

49. *Presstime*, "Partner Power and the Next Media Age," 39. Statement attributed to Susan Decker, a newspaper industry analyst with the brokerage firm Donaldson, Lufkin & Jenrette.

50. *Presstime*, "Panel's Advice: Get in Gear; Content, Identity Drive Development," 27.

51. Potter, "A Guide to Online Avenues," S1.

52. Ibid., S1–S2. Statement attributed to Kenneth Paulson, executive editor and vice president for news at Gannett's suburban newspaper division.

53. Ibid., S7.

54. Ibid., S9–S10.

55. Ibid., S11.

56. Ibid., S12.

57. Ibid., S4.

58. Ibid., S9.

59. McKenna, "The Future Is Now," par. 27. Statement attributed to Victor Perry, director of new business development for the *Los Angeles Times*.

60. Potter, "Joe Reader Meets the E-Paper: You've Read about Them. We Decided to Actually Read Them," 44–45.

61. Ibid.

62. Potter, "A Guide to Online Avenues," S8.

63. America Online Inc., "Digital City, Inc., to Launch Four New 'Virtual Cities.'"

64. Moeller, "The High-Tech Trib," par. 56.

65. Potter, "A Guide to Online Avenues," S9. Statement attributed to Eugene Quinn, general manager of Tribune Interactive Network Services.

66. Ibid., S7. Statement attributed to Henry Scott, vice president for new media and new products at the New York Times Co.

67. Ibid., S8.

68. Ibid. Statement attributed to Bill Mitchell, head of Mercury Center, the online information services division of the *San Jose Mercury News*.

69. Markoff, "17 Companies in Electronic News Venture," pars. 14–15.

70. Ibid. Statement attributed to Robert Ingle, executive editor of the *San Jose Mercury News*.

71. America Online, "America Online Becomes First Billion Dollar Interactive Services Company," pars. 1, 5.

72. Carter, "Online Advertising," 38–39. Statement attributed to Frederick J. Tuccillo, director of electronic publishing for *Newsday*.

73. Ibid., 40. Statement attributed to Ted Leonis, president of AOL Services Co.

74. Ibid., 38. Statement attributed to Erica Gruen, senior vice president with the New York advertising agency Saatchi & Saatchi.

75. Shedden, "New Media Timeline (1988)."

76. Shapiro, "The Media Business; New Features Are Planned by Prodigy," par. 6.

77. Shedden, "New Media Timeline (1988)."

78. Conhaim, "Developing Videotex as a Consumer Medium," par. 4.

79. Ibid., par. 3.

80. Ibid., par. 12.

81. Potter, "Cox Seeks Power in Partnerships," 19. Statement attributed to David Easterly, president of Cox Newspapers.

82. Ibid.

83. Potter, "A Guide to Online Avenues," S4.

84. Ibid. Statement attributed to Michael T. Darcy, Prodigy's communications manager.

85. *Presstime*, "Atlanta's Custom Connection," 8.

86. *Presstime*, "Partner Power and the Next Media Age," 38.

87. Gipson, "Wiring the Waltons," 16. Statements attributed to Victor Perry, vice president of new business development at the *Los Angeles Times*.

88. Potter, "A Guide to Online Avenues," S7. Statement attributed to Carl Crothers, deputy managing editor for the *Tampa Tribune* and team leader for its Tampa Bay Online.

89. Davis, "Profiles: Frank A. Daniels III; George Schlukbier," 24–25.

90. Gipson, "New Media: Newspapers and Related Organizations Continue to Launch On-Line Services," 13.

91. Carter, "Online Advertising," 39. Statement attributed to Eugene Quinn, general manager of Tribune Interactive Network Services.

92. Reid, *Architects of the Web*, 14.

93. Ibid.

94. Taylor, "AOL: Won't Tangle with Web," 6. Statement attributed to David Gang, vice president of product marketing for America Online.

95. Boczkowski, *Digitizing the News*, 48.

96. Ibid.

97. Fink, *Strategic Newspaper Management*, 26.

98. McKenna, "The Future Is Now," pars. 9, 11.

5. The Emerging Internet Threatens the Established Publishing Model

1. Wolfe, "The (Second Phase of the) Revolution Has Begun," par. 1.
2. Microsoft, Microsoft Announces Availability of the Microsoft Network; Consumers Worldwide Invited to Get Online with Windows 95 and MSN, par. 1.
3. Rainie and Bell, "The Numbers That Count," 44.
4. The *Merriam-Webster Dictionary* reported the first known use of this play on the word literati occurred in 1992 and was applied to "persons well-versed in computer use and technology." (http://www.merriam-webster.com/dictionary)
5. Kapor, "Where Is the Digital Highway Really Heading? The Case for a Jeffersonian Information Policy," pars. 3–4.
6. Arnold and Arnold, "Vectors of Change: Electronic Information from 1977 to 2007," par. 56.
7. Osder, "The Little Browser That Could and the New Media Revolution," par. 1.
8. Yamamoto, "Legacy: A Brave New World Wide Web," par. 5.
9. Carlson, "Online Timeline," 69.
10. Markoff, "Business Technology: A Free and Simple Computer Link," par. 1.
11. Ibid., par. 3.
12. Reid, *Architects of the Web: 1,000 Days That Built the Future of Business,* 2.
13. Ibid.
14. Ibid.
15. Ibid., 44.
16. *Presstime*, "The Content Kings' Conundrum," 27. Statement attributed to Kathleen Criner, an industry consultant who had been senior vice president for industry development at the NAA.
17. Chan-Olmsted, "Mergers, Acquisitions, and Convergence: The Strategic Alliances of Broadcasting, Cable Television, and Telephone Services," 41.
18. Ibid., 33–34.
19. Boczkowski, *Digitizing the News: Innovation in Online Newspapers,* 37.
20. *Presstime*, "Future Bound: Products, Systems and Practices Preview What's Next," 31. Statement attributed to Cathleen Black, president and chief executive officer of the NAA.
21. *Presstime*, "Putting It Together: The Las Vegas Trade Show Proved a Technological Oasis of Shared Ideas, Strategies and Solutions," 31. Statement attributed to Charles T. Brumback, chairman and chief executive officer of Tribune Co. and a chairman of the NAA.
22. *Presstime*, "Future Bound: Products, Systems and Practices Preview What's Next," 31. Statement attributed to Uzal Martz, publisher of the *Pottsville Republican* in Pennsylvania and a chairman of the NAA.
23. *Presstime*, "Association Now at Home on the Web," 6.

24. Shaw, "Videotext, the Sequel," 14.
25. Shedden, "New Media Timeline (1995)."
26. Shedden, "New Media Timeline (1996)."
27. Peng, Tham, and Xiaoming, "Trends in Online Newspapers: A Look at the U.S. Web," par. 4.
28. Knight-Ridder Inc. "Knight-Ridder Closes Information Design Laboratory," par. 2.
29. Boczkowski, *Digitizing the News*, 48.
30. Peng, Tham, and Xiaoming, "Trends in Online Newspapers," par. 8.
31. Lee, Ya-Ching, "Newspaper Online Services: A Successful Business?; Lessons Learned from Videotext Failure," 11.
32. Ibid., 23.
33. Lewis, "The Media Business: Big Newspapers to Help Locals on Internet," par. 3.
34. *New York Times*, "Information Technology: The Times Is Joining New On-Line Service," par. 4.
35. Ibid., par. 5.
36. Toner, "Standards Bearers Enter the Web," 76.
37. Ibid. Statement attributed to Robert Ingle, vice president of Knight-Ridder's new media operations and the company's NCN representative.
38. Lewis, "The Media Business," par. 5. Statement attributed to John S. Reidy, a media analyst for Smith Barney, Inc.
39. Toner, "Standards Bearers Enter the Web," 76.
40. Lewis, "The Media Business," par. 7.
41. Ibid., par. 16.
42. *Presstime*, "Interactivity: On-line's Edge," 50. Peter Winter was vice president of market development for Cox Newspapers and the interim chief executive officer for NCN.
43. Ibid.
44. Peterson, "New Service Skims 150 Newspapers for Its Users," par. 1.
45. Ibid., par. 2.
46. *Presstime*, "Six Papers Blaze CareerPath," 15.
47. Ibid.
48. Gipson, "New Media: Strategic Moves," 12.
49. Wronski, "Media Companies Seeking New Turf," 8.
50. Pelline, "Print to Try Packaging with NewsWorks," par. 5. Statement attributed to John Papanek, NCN's editor-in-chief.
51. Ibid., par. 14.
52. Ibid., par. 7.
53. Ibid., par. 15.
54. Peterson, "New Service Skims 150 Newspapers for Its Users," par. 6. Statement attributed to Steve Mitra, an analyst with Jupiter Communications.
55. O'Leary, "NewsWorks Brings New Depth to Web News," par. 10.

56. Knight-Ridder Inc., "Knight-Ridder Inc. Agrees to Purchase Dialog Information Services, Inc. from Lockheed Corp. for $353 million."

57. O'Leary, "NewsWorks Brings New Depth to Web News," pars. 18–19.

58. Cole, David M., "Industry Saddened, Chastened by Demise of NCN," par. 16.

59. Miles, "New Century Networks Regroups," pars. 1, 5.

60. Miles, "News Megasite Cuts Content, Staff," par. 8.

61. Cole, David M. "Industry Saddened, Chastened by Demise of NCN," par. 5.

62. Yelvington, "Nostalgia for the New Century Network," par. 1

63. Ibid., pars. 3–4.

64. Dugan, "New-Media Meltdown at New Century: How a Big Online Newspaper Venture Bit the Dust," par. 2.

65. Ibid., par. 3.

66. Ibid., par. 2. Statement attributed to David Morgan, president of Real Media Inc.'s online advertising agency.

67. Pelline, "Print to Try Packaging with NewsWorks," par. 15.

68. Peterson, "New Service Skims 150 Newspapers for Its Users," par. 5.

69. Dugan, "New-Media Meltdown at New Century," par. 9.

70. Ibid.

71. Ibid., pars. 4, 9, 14.

72. Cole, David M., "Industry Saddened, Chastened by Demise of NCN," par. 2.

73. Dugan, "New-Media Meltdown at New Century," par. 7. Statement attributed to Al Sikes, a former commissioner of the FCC who was president of Hearst New Media.

74. Ibid., par. 6. Statement attributed to Harry Chandler, who ran the new media operations for the *Los Angeles Times.*

75. Ibid., par. 12.

76. Ibid., par. 3. Statement attributed to Peter Winter, an executive with Cox Enterprises who had served as NCN's interim chief executive officer at its creation.

77. Shedden, "New Media Timeline (1995)."

78. *Atlanta Journal-Constitution,* "New Cox Unit to Focus on the Internet," D1.

79. Pelline, "Cox Muscles into Local Markets," par. 3.

80. Snyder and Kerwin, "Publishers Fight for Local Dollars in Online Space: Newspapers Get More Aggressive with Web Sites," par. 6.

81. New York Times Co., "Newspaper Establishes Internet Presence."

82. Ibid. Statement attributed to Arthur Sulzberger, Jr., publisher of the *New York Times.*

83. Toner and Gipson, "New Media: On Line," 10.

84. Goldstein, "A Few Veterans Look Back: Considering the Early Years of Internet Journalism," par. 16. Statement attributed to Martin Nisenholtz, chief executive officer of *New York Times* Digital.

85. New York Times Co., "Newspaper Establishes Internet Presence."

86. Aikat, "News on the Web: Usage Trends of an On-Line Newspaper," 100.

87. Gipson, "Behold the Megawebsite," 11.

88. Ibid.

89. Toner and Gipson, "New Media: On Line," 10.

90. Ibid.

91. Picard, "Changing Business Models of Online Content Services: Their Implications for Multimedia and Other Content Producers," 62.

92. Hooker, "A Library on My Belt," 49. Michael Hooker was president of the University of Massachusetts.

93. Ibid.

94. *Presstime*, "Tomorrow Today," 46. The speaker was W. Russell Neuman, a research fellow at the Massachusetts Institute of Technology Media Laboratory.

95. Lapham, "The Evolution of the Newspaper of the Future," par. 9.

96. Ingle, "Newspaper vs. On-Line Versions: A Discussion of the Old and New Media," par. 22.

97. Chyi and Sylvie, "Competing with Whom? Where? And How? A Structural Analysis of the Electronic Newspaper Market," 5.

98. Ibid.

99. Cole, David M., "Planner: The Virtual Library; Yesterday's Paper Will Provide Tomorrow's Revenue through Digital Access," 47.

100. Ibid., 47–48.

101. Ibid., 52. Statement attributed to George Schlukbier, a professional librarian and vice president and editor of NandO.net, the Internet operations of the *News & Observer* in Raleigh, North Carolina.

102. Shaw, "Videotext, the Sequel," 14. Uzal Martz was president/publisher of the *Pottsville Republican* in Pennsylvania and a vice-chairman of the NAA at the time of these remarks.

103. Cole, David M. "Webservations," 69.

104. Chyi and Sylvie, "Competing with Whom? Where? And How?"13.

105. Chyi and Sylvie, "Online Newspapers in the U.S.: Perceptions of Markets, Products, Revenue and Competition," 75.

106. Ibid.

107. Fulton, "A Tour of Our Uncertain Future," 20.

108. Zuckerman, "Newspapers Balk at Scooping Themselves on Their Own Web Sites," par. 6.

109. Ibid., par. 7. Statement attributed to Bob Ryan, a director at Knight-Ridder Inc.'s online operations known as Mercury Center.

110. Wright, Donald K., "The Magic Communication Machine: Examining the Internet's Impact on Public Relations, Journalism, and the Public," 20. Statistics attributed to *The Seventh Annual Middleberg/Ross Survey of Media in the Wired World 2000* (New York: Middleberg, 2001).

111. Leckner, "Is the Medium the Message?: The Impact of Digital Media on the Newspaper Concept," 77.

112. Lee, William Chee-Leong, "Clash of the Titans: Impact of Convergence and Divergence on Digital Media," 109.

113. Ibid., 86.

114. Katz, "Online or Not, Newspapers Suck," par. 51. Jon Katz is a media critic.

115. Ibid., par. 46.

116. Schlukbier, "Internet: Newspapers' Best Strategy?" S15. Schlukbier was director of new media for the *News & Observer* in Raleigh, North. Carolina.

117. Ibid.

118. Hollander, "Talk Radio, Videotext and the Information Superhighway," 60. Hollander is a journalism professor at the University of Georgia.

119. Lapham, "The Evolution of the Newspaper of the Future," par. 30.

120. Pavlik, "The Impact of Technology on Journalism," 235.

121. Ibid.

122. Ibid., 236.

123. Ibid., 235. Pavlik attributed the agenda-setting concept to Bernard Cohen (1963), a political scientist quoted as stating: "The press may not be successful much of the time in telling us what to think, but it is stunningly successful in telling us what to think about."

124. Pavlik, "The Future of Online Journalism: Bonanza or Black Hole?" 30.

125. Mings and White, "Profiting from Online News: The Search for Viable Business Models," 70.

126. The origin of the phrase "The Daily Me" is often attributed to work from the Media Laboratory at MIT.

127. Mings and White, "Profiting from Online News," 68–70.

128. Bender, et al., "Enriching Communities: Harbingers of News in the Future," 379.

129. Kilker, "Shaping Convergence Media," 20.

130. Gipson, "Bill's On-Line AdVenture," 31.

131. Toner, "Competing for Cyberturf," par. 1.

132. Ibid., par. 6.

133. Ibid., par. 11. Statement attributed to Ellen Siminoff, director of Yahoo Communities.

134. Ibid., par. 4.

135. Caruso, "Show Me the Money!" par. 15. Denise Caruso wrote a column on digital commerce for the *New York Times*; was a visiting lecturer on interactive media at Stanford University; and a visiting scholar at Interval Research, an industry think tank.

136. Fine, "The Daily Paper of Tomorrow: It Won't Look the Same. But with Reimagining, the Local Daily Ain't Dead Yet," par. 2. Statement attributed to Mark Colodny, managing director of Warburg Pincus.

137. Hinton, "Portal Sites: Emerging Structures for Internet Control," 17.

138. Shaw, "Portals: An Introduction," 2.

139. Hinton, "Portal Sites," 17.

140. Mansell, "New Media Competition and Access: The Scarcity-Abundance Dialectic," 160.

141. Shaw, "Portals: An Introduction," 2.

142. Damsgaard, "Managing an Internet Portal," 408.

143. Mansell, "New Media Competition and Access," 167.

144. Hinton, "Portal Sites," 29.

145. Schonfeld, "The Flickrization of Yahoo," 158. Statement attributed to Terry Semel, chief executive of Yahoo.

146. Meisel and Sullivan, "Portals: The New Media Companies," 477–486.

147. Chyi and Sylvie, "Competing with Whom? Where? And How?," 17.

148. Toner, "Competing for Cyberturf," par. 10.

149. Newspaper Association of America.

150. *Presstime*, "Convention Report: A Classified Wake-Up Call," 18.

151. Ibid.

152. Ibid.

153. Burroughs, "Crisis in Classifieds," 39–40.

154. Mings and White, "Profiting from Online News," 76.

155. Peng, Tham, and Xiaoming, "Trends in Online Newspapers," par. 55.

156. Mings and White, "Profiting from Online News," 76. This research referenced an "Online Classifieds Report" produced by *Editor & Publisher*.

157. Ibid., 77.

158. Ibid., 78. Projection attributed to Jupiter Communications.

159. Romenesko, "What the Near Future Holds," par. 7. Statement attributed to Mary Modahl, an analyst with Forrester Research.

160. Burroughs, "Crisis in Classifieds," 39–40. Statement was attributed to James Williams, an advertising consultant.

161. Wyman, "The Last Link: Employment Listings Are Another Forfeited Franchise," par. 4.

162. Wetmore, "Classified Market Heats Up with New AOL Player: NAA Seeks a Standard to Build on Web Strengths; More Buy into Classified Ventures," par. 15. Statement attributed to Eric Wolferman, senior vice president for technology at the NAA.

163. Wetmore, "Classified Ventures: What NCN Taught a Newcomer; On-Line Enterprise Is Owned by Three Media Companies Whose Fingers are at Arm's Length," par. 8. Statement attributed to Tim Landon of Classified Ventures.

164. Cole, D., "NAA's Ad Numbers May Be Signaling Beginning of the End," par. 4.

165. Ibid., par. 3.

166. Newspaper Association of America.

167. Lacy and Martin, "Competition, Circulation and Advertising," 30.

168. Ibid., 34.

169. Flanigan, "Newspapers Aren't Dying, They're Just Turning a Page," C1.

170. Ibid.

171. Groves, "1996 Forecast: A Look at What Was, Why and What May Be," 26.

172. Maxwell and Wanta, "Advertising Agencies Reduce Reliance on Newspaper Ads," par. 8.

173. Boczkowski, *Digitizing the News*, 67.

174. Ibid.

175. Caruso, "Show Me the Money!" pars. 3–4.

176. Fink, *Strategic Newspaper Management*, 228.

177. Peng, Tham, and Xiaoming, "Trends in Online Newspapers," par. 56.

178. Ibid.

179. Lee, Ya-Ching "Newspaper Online Services," 7.

180. Chyi and Sylvie, "Online Newspapers in the U.S.," 73.

181. Gipson, "@ Your Name Here," 11.

182. Ibid.

183. Chyi and Sylvie, "Online Newspapers in the U.S.," 73.

184. McMillan, "Who Pays for Content? Funding in Interactive Media," par. 65.

185. Washington, "Paper Battles Online Services," par. 3. Statement attributed to Ron Mann, director of automotive advertising for the *Los Angeles Times*.

186. Newspaper Association of America.

187. Maddox, "Joint Report Reveals Online Ad Recovery; Traditional Advertisers Embrace Rich Media, but Not All Sectors Benefitting," par. 3.

188. Boczkowski, *Digitizing the News*, 178.

189. Ibid., 71.

190. Ibid.

191. Ibid., 67.

6. Mergers, Convergence, and an Industry under Siege

1. Gershon and Alhassan, "AOL Time Warner & WorldCom Inc.: Corporate Governance and Diffusion of Authority," 8.

2. Miller, "The Economy's New Clothes?" 371.

3. Jones and Marx, "Weaned on Crisis, Landing on Top: The Man Behind AOL Is an Online Visionary Who's Had His Share of Crashes," par. 9.

4. Bontis and Mill, "Dot-Bomb Post-Mortem: Web-Based Metrics and Internet Stock Prices," 17.

5. *Yahoo Finance*, "NASDAQ Composite Interactive Chart."

6. Gershon and Alhassan, "AOL Time Warner & WorldCom Inc.," 8.

7. Hickey, "Coping with Mega-Mergers," 16.

8. Ibid., 17.
9. Wolf, "Media Mergers: The Wave Rolls On," par. 5. Michael Wolf was a media industry analyst for McKinsey & Co.
10. Ibid., par. 2.
11. Ibid., par. 3.
12. Fancher, "Disturbing Findings from New Survey," par. 10.
13. Morton, "The Emergence of Convergence," 88.
14. Fuller, "Merging Media to Create an Interactive Market: New Strategies Are Used to Fund the Expensive Business of Newsgathering," 33.
15. Elder, "The Return of the Dead-Tree Media," par. 6.
16. Fuller, "Merging Media to Create an Interactive Market," 33.
17. Ibid.
18. Elder, "The Return of the Dead-Tree Media," pars. 1, 5.
19. Gunther, "Publish or Perish?," par. 6.
20. Cox Enterprises, *CIMCities Unit Option Plan*, 4.
21. Gipson, "Playing the Online Game to Win," 27.
22. *Yahoo Finance*, "NASDAQ Composite Interactive Chart."
23. Bontis and Mill, "Dot-Bomb Post-Mortem," 19.
24. Ibid.
25. Cunningham, "Rush to the Web—and Away from It: New Media Were All So Exciting and Breathtaking. And Then Came the Crash," 120.
26. Fost, "Rumors of Web's Demise Greatly Exaggerated: Old Media's Retreat from Internet Spending Doesn't Signify Surrender," par. 13. Statement attributed to Dan Finnigan, president of KnightRidder.com.
27. Gershon and Alhassan, "AOL Time Warner & WorldCom Inc.," 2.
28. Svaldi, "Online Outlets Link Newspapers, Profits; Publisher Tells of Internet Boost," par. 4. Statement attributed to William Dean Singleton, publisher of the *Denver Post*.
29. Gipson, "Playing the Online Game to Win," 31.
30. Svaldi, "Online Outlets Link Newspapers, Profits," par. 6. The industry analysis was attributed to Rusty Coats, director of new media at MORI Research.
31. Gipson, "Playing the Online Game to Win," 28.
32. Ibid.
33. Nerone and Barnhurst, "Beyond Modernism: Digital Design, Americanization and the Future of Newspaper Form," 477.
34. Gipson, "Playing the Online Game to Win," 28. Statements attributed to Martin Nisenholtz, chief executive officer of New York Times Digital.
35. Ibid., 27. Statements attributed to Chris Jennewein, Internet director for the *San Diego Union-Tribune*.
36. Gunther, "Publish or Perish?," par. 4.
37. Hickey, "Coping with Mega-Mergers," 16. Statement attributed to James Fallows, a journalist and media critic who had written *Breaking the News* in 1996.

38. Dowling, Lechner, and Thielmann, "Convergence—Innovation and Change of Market Structures between Television and Online Services," 31–35.

39. Thielmann and Dowling, "Convergence and Innovation Strategy for Service Provision in Emerging Web-TV Markets," 4.

40. Ibid.

41. Dominick, *The Dynamics of Mass Communication*, 20.

42. Hickey, "Coping with Mega-Mergers," 20.

43. Ibid.

44. Scott, "A Contemporary History of Digital Journalism," 93.

45. Dominick, *The Dynamics of Mass Communication*, 21.

46. Scott, "A Contemporary History of Digital Journalism," 101.

47. Singer, "Strange Bedfellows? The Diffusion of Convergence in Four News Organizations," 4.

48. Ibid., 3.

49. Lawson-Borders, "Integrating New Media and Old Media: Seven Observations of Convergence as a Strategy for Best Practices in Media Organizations," 92.

50. Wirtz, "Convergence Processes, Value Constellations and Integration Strategies in the Multimedia Business," 15.

51. Ibid., 22.

52. Wendland, "Convergence: Repurposing Journalism," par. 9. Estimate attributed to James Gentry, dean of the School of Journalism at the University of Kansas.

53. Ibid., pars. 11, 23.

54. duPlessis and Li, "Cross-Media Partnership and Its Effect on Technological Convergence of Online News Content: A Content Analysis of 100 Internet Newspapers," 166.

55. Singer, "Strange Bedfellows?," 4.

56. Ibid., 3.

57. Ibid., 17.

58. Lawson-Borders, "Integrating New Media and Old Media," 94.

59. Ibid., 92.

60. Ibid.

61. Montagne, "Part Two of the Series on Media Convergence Looks at a Media Empire in Lawrence, Kansas," par. 10.

62. Ketterer, et al., "Case Study Shows Limited Benefits of Convergence," 52, 55.

63. Ibid., 53.

64. Dotinga, "An Urge to Converge," 9. Statement attributed to Kelly Dyer, general manager of NewsOK.com.

65. Wirtz, "Convergence Processes, Value Constellations and Integration Strategies," 15.

66. Chon, et al., "A Structural Analysis of Media Convergence: Cross-Industry Mergers and Acquisitions in the Information Industries," 142.

67. Kolodzy, "Everything That Rises," 61. Janet Kolodzy was a journalism professor at Emerson College.

68. Colon, "The Multimedia Newsroom," par. 11.

69. Wendland, "Convergence: Repurposing Journalism," pars. 1–5. Mike Wendland was a fellow at the Poynter Institute, a newspaper industry research and training center.

70. Thelen, "Tampa's Convergence Lessons," 7. Gil Thelen was executive editor of the *Tampa Tribune* and a vice president with Media General.

71. Ibid.

72. Tompkins and Colon, "Tampa's Media Trio," 46.

73. Colon, "The Multimedia Newsroom," par. 25.

74. Singer, "Strange Bedfellows?," 10.

75. Thelen, "Tampa's Convergence Lessons," 8.

76. Kolodzy, "Everything That Rises," 61.

77. Tompkins and Colon, "Tampa's Media Trio," 52.

78. Ibid., 53.

79. Thelen, "Tampa's Convergence Lessons," 8.

80. Lawson-Borders, "Integrating New Media and Old Media," 94.

81. Ibid.

82. Colon, "The Multimedia Newsroom," par. 23.

83. Ibid., par. 17.

84. Colon, "The Multimedia Newsroom," par. 20.

85. Thelen, "Tampa's Convergence Lessons," 9.

86. Ibid., 7.

87. Carr, "Searching for TV-Newspaper Convergence: Studies Show Little Cooperation in Station-Newspaper Partnerships," par. 1.

88. Glaser, "Annals of Integration: New York Times and Times Digital," par. 19.

89. Gordon, "The Meanings and Implications of Convergence," 65.

90. Scott, "A Contemporary History of Digital Journalism," 113–114.

91. Singer, "Strange Bedfellows?," 13.

92. Colon, "The Multimedia Newsroom," par. 20.

93. Carr, "Searching for TV-Newspaper Convergence," par. 14. Observation attributed to Debora Halpern Wenger, a Virginia Commonwealth University journalism professor who had been an assistant news director at Tampa's WFLA.

94. Ketterer, et al, "Case Study Shows Limited Benefits of Convergence." 61.

95. Carr, "Searching for TV-Newspaper Convergence," par. 5.

96. Finkle, "Cox Interactive's Orange County, Calif., Web Portal to Shut Down," par. 6. Statement attributed to Scott Whiteside, then chief operating officer of Cox Interactive Media.

97. Sullivan, "Cox Shifts Web Strategy to Local Control," par. 11. Statement attributed to Scott Whiteside, then chief operating officer of Cox Interactive Media.

98. Smolkin, "Adapt or Die," 17.

99. Glaser, "Annals of Integration: New York Times and Times Digital," par. 1.

100. Ibid., par. 16. Statement attributed to Denise Warren, chief advertising officer for the New York Times Co.

101. Lind, "Convergence: History of Term Usage and Lessons for Firm Strategists," 1.

102. Ibid.

103. Ibid., 11.

104. Jung, "The Bigger, the Better? Measuring the Financial Health of Media Firms," 246.

105. Ibid.

106. Rothenberg, "Convergence Likely to Render Restriction of Liquor Ads Futile," 14.

107. Bowman and Willis, "The Future Is Here, But Do News Media Companies See It?," par. 2.

108. Zhang, "Online Newspapers Readership Increases," par. 3. Statement attributed to Gerry Davison, senior media analyst with Nielsen/NetRatings.

109. Ibid., pars. 1–2.

110. Mann, "Deal Fuels Debate on Future of Print," D4. Statement attributed to James P. Rutherfurd, executive vice president of Veronis, Suhler Stevenson, a private equity firm that specializes in the media industry.

111. Ibid. Statement also attributed to Rutherfurd.

112. Silverthorne, "Read All About It! Newspapers Lose Web War: Q&A with Clark Gilbert," par. 10.

113. Ibid., par. 11.

114. Smolkin, "Adapt or Die," 18.

115. Newspaper Association of America.

116. Google, "Google Announces Record Revenues for Fourth Quarter and Fiscal Year 2004," par. 18.

117. Gillette, "Newspapers, Television Finding Ways to Make Websites Pay," par. 14.

118. Battelle, "The Wizard of Ads," 120. Statement attributed to Omid Kordestani, senior vice president for global sales and business development for Google.

119. Ibid.

120. Gunther, "Publish or Perish?," par. 32. Statement attributed to Mark Willes, chief executive officer of Times Mirror.

121. Thompson and Wassmuth, "Few Newspapers Use Online Classified Interactive Features," 26.

122. Ubinas and Yang, "Classified Ads: How Newspapers Can Fight Back," par. 2.

123. Silverthorne, "Read All About It! Newspapers Lose Web War," par. 24.

124. Gipson, "Playing the Online Game to Win," 27.

125. Gunther, "Publish or Perish?" par. 29. Statement attributed to Jeff Taylor, chief executive officer of Monster.com.

126. Smolkin, "Adapt or Die," 21. Statement attributed to media analyst Gordon Borrell.

127. Knowledge@Wharton, "All the News That's Fit to . . . Aggregate, Download, Blog: Are Newspapers Yesterday's News?" par. 7. Statement attributed to Lawrence Hrebiniak, a professor of management at the Wharton School of the University of Pennsylvania.

128. Steinberg, "Newspaper Woes Are Black and White," B3.

129. Ibid.

130. Ibid. Statement was attributed to advertising consultant David Cross.

131. Liedtke, "Web Growth, Innovation Threatens Papers," pars. 20–21.

132. Lehman-Wilzig and Cohen-Avigdor, "The Natural Life Cycle of New Media Evolution," 707.

133. Ibid., 718–719.

134. Edmonds, "Newspapers: Economics," par. 2.

135. Gunther, "Publish or Perish?," par. 29. Statement attributed to David Israel, CEO of Classified Ventures, a company backed by newspaper industry investors that focused on online automotive and apartment advertising.

136. Smolkin, "Adapt or Die," 18. Statement attributed to Jay Smith, president of Cox Newspapers.

137. Edmonds, "Newspapers: Economics," par. 10.

138. Smolkin, "Adapt or Die," par. 17.

139. Margolies, "Chain's Sale Is Milestone for Industry, Star," A1.

140. Liedtke, "Knight-Ridder's Demise Reflects Sobering Times for U.S. Newspapers," pars. 10–11. Statistics attributed to John Morton, an industry analyst.

141. Chaffin, "Knight-Ridder Investor Pushes for Company Sale," pars. 2, 8.

142. Ibid., pars. 5, 7.

143. Knight-Ridder, "Knight-Ridder Announces Exploration of Strategic Alternatives," par. 2.

144. Rosenthal, "Media Column," par. 7.

145. Seelye, "What-Ifs of a Media Eclipse," par. 6. Statement attributed to Jay T. Harris, a former publisher for Knight-Ridder at the *San Jose Mercury News*.

146. Ibid., par. 19.

147. Hill, "Sale Would End Family's Empire After 114 Years: P. Anthony Ridder, Pushed by Investors to Take Bids, Would Be Out as CEO. The Ridders Bought Their First Newspaper in 1892," par. 6.

148. Seelye, "What-Ifs of a Media Eclipse," par. 18.

149. Kasler, "McClatchy Purchase Heralds New Era: CEO Pruitt Says Buying Knight-Ridder Is the Right Move Despite High Stakes," par. 6. Statement attributed to Gary Pruitt, chairman and chief executive officer of McClatchy Co.

150. Mann, "Deal Fuels Debate on Future of Print," D4.

151. Van Riper, "Ink-Stained Wretches," par. 9. Statement attributed to Murray D. Schwartz, a mergers and acquisitions partner with Katten Muchin Rosenman, which specialized in newspaper industry transactions.

152. Weisenthal, "After Layoffs at McClatchy, a Focus on Pruitt's Pay; What Are McClatchy's Digital Goals?" par. 3.

153. Smolkin, "Adapt or Die," 18. Statement attributed to Stephen Gray, managing director of Newspaper Next, a program of the American Press Institute.

154. Conhaim, "Newspapers in Transition, Part 2: Envisioning the Future of Newspapers," par. 1.

155. Mann, "Deal Fuels Debate on Future of Print," D4. Statement attributed to Dean Mills, dean of the school of journalism at the University of Missouri.

156. Smolkin, "Adapt or Die," 20. Statements attributed to Conrad Fink, a journalism professor at the University of Georgia and author of *Strategic Newspaper Management*.

157. Seelye, "What-Ifs of a Media Eclipse," par. 17. Statement attributed to James M. Naughton, former executive editor of the *Philadelphia Inquirer*, which had been a Knight-Ridder newspaper. Naughton was also a former president of the Poynter Institute for Media Studies.

158. Picard, "Capital Crisis in the Profitable Newspaper Industry," 11.

159. Ibid., 12.

160. Ibid.

161. Liedtke, "Web Growth, Innovation Threatens Papers," par. 16.

162. Rosen, "Laying the Newspaper Gently Down to Die," par. 14.

163. Picard, "Capital Crisis in the Profitable Newspaper Industry," 12.

164. Seelye, "What-Ifs of a Media Eclipse," par. 40. Statement attributed to Lauren Rich Fine, a media industry analyst with Merrill Lynch.

165. Ibid., par. 47. Statement attributed to Anthony Ridder, the former chief executive officer of Knight-Ridder.

166. Picard, "Capital Crisis in the Profitable Newspaper Industry," 11.

7. Connecting the Lessons of History

1. Adams, "Gannett Revamps USA Today for Web, Pares Work Force," B5.

2. Ibid.

3. *Wired*, "The Web Is Dead," 118.

4. Anderson, "Who's to Blame: Us," 123–124.

5. *Wireless News*, "RTG Ventures Enters Mobile App Market," par. 2.

6. eMarketer, "Mobile Passes Print in Time-Spent Among U.S. Adults," par. 6.

7. Ibid.

8. Albarran, *The Media Economy*, 59.

9. Sterling, "Digital First: What Does It Mean, And Where Will It Take Us?" par. 2.

10. Palser, "The Ins and Outs of iPad Apps," par. 1.

11. Ives, "What Matters Most in Magazine and Newspaper iPad Apps? Quality, Report Says," par. 6.

12. Palser, par. 3.
13. Underwood, "Reinventing the Media," 27.
14. Newspaper Association of America, "Trends and Numbers: Advertising Expenditures."
15. Wall Street Equity Research, " Analyst Study on Gannett and New York Times—The End of an Era and the Start of a New Generation of E-Newspapers," par. 6.
16. *The Economist*, "Another Brick in the Wall," par. 5.
17. Ibid.
18. Associated Press, "Analyst: Digital Fees Not Enough to Help NY Times."
19. Ibid.
20. Udell, *The Economics of the American Newspaper*, 14.
21. Barrett and Siegel, "Time for Newspaper Folks to Fight Back! Here's How," par. 6.
22. Jones, "The Bias of the Web," 181.
23. Ibid.
24. Ibid.
25. Knowledge@Wharton, "All the News That's Fit to . . . Aggregate, Download, Blog: Are Newspapers Yesterday's News?" par. 39. Statement attributed to Michele Weldon, an assistant professor at Northwestern University's Medill School of Journalism.
26. Udell, *The Economics of the American Newspaper*, 24.
27. Schonfeld, "Tuning Up Big Media: A Modest Proposal for Saving Time Warner and the Entire Industry from Themselves," 62.
28. Bryant, "Message Features and Entertainment Effects," 254.
29. Boczkowski, "The Mutual Shaping of Technology and Society in Videotex Newspapers," 263.
30. Compaine, *The Newspaper Industry in the 1980s: An Assessment of Economics and Technology*, 213.
31. Gunther, "Publish or Perish?" par. 29. Statement attributed to David Israel, CEO of Classified Ventures, a company backed by newspaper industry investors that focused on online automotive and apartment advertising.
32. Rosen, "Laying the Newspaper Gently Down to Die," par. 14.
33. Kirchhoff, "The U.S. Newspaper Industry in Transition," 7.
34. Knowledge@Wharton. "All the News That's Fit to . . . Aggregate, Download, Blog: Are Newspapers Yesterday's News?" par. 6. Statement attributed to Wharton management professor Lawrence Hrebiniak.
35. *Presstime*, "Putting It Together: The Las Vegas Trade Show Proved a Technological Oasis of Shared Ideas, Strategies and Solutions," 31. Statement attributed to Charles T. Brumback, chairman and chief executive officer of Tribune Co. and chairman of the NAA.
36. Sparks, *Media Effects Research*, 240.

37. Fine, "The Daily Paper of Tomorrow: It Won't Look the Same. But with Reimagining, the Local Daily Ain't Dead Yet," par. 2. Statement attributed to Mark Colodny, managing director of Warburg Pincus.
38. Allen, "Old Media, New Media, Not Media?" par. 2.
39. Schonfeld, "The Flickrization of Yahoo," 158. Statement attributed to Terry Semel, chief executive of Yahoo.
40. Dominick, *The Dynamics of Mass Communication,* 19.

Bibliography

Adams, Russell. "Gannett Revamps USA Today for Web, Pares Work Force." *Wall Street Journal*, August 28–29, 2010: B5.

Aikat, Debashis. "News on the Web: Usage Trends of an On-Line Newspaper." *Convergence* 4, no. 94 (1998): 94–110.

Albarran, Alan B. *The Media Economy*. New York: Routledge, 2010.

Alber, Antone F. *Videotex/Teletext: Principles and Practices*. New York: McGraw-Hill, 1985.

Allen, Matthew. "Old Media, New Media, Not Media?" *Net Critic*, October 9, 2009. http://www.netcrit.net/ideas/old-media-new-media-not-media/.

America, Anna. "RBOCs Battle Enters New Phase." *Presstime*, August 1991: 45.

America Online Inc. "America Online Becomes First Billion Dollar Interactive Services Company." August 8, 1996, press release. http://www.thefreelibrary.com/america+online+becomes+first+billion+dollar+interactive+services . . . -a018561554.

———. Digital City, Inc., to Launch Four New 'Virtual Cities.'" August 15, 1996, press release. http://www.highbeam.com/doc/1G1-18578809.html.

American Telephone and Telegraph Company. "Directory and Public Services." *American Telephone and Telegraph Company 1979 Annual Report*. February 8, 1980: 12. http://www.porticus.org/bell/att/1979/att_1979.htm.

Anderson, Chris. "Who's to Blame: Us." *Wired*, September 2010, 123–127; 164.

Arnold, Stephen E., and Erik S. Arnold. "Vectors of Change: Electronic Information from 1977 to 2007." *Online*, July 1, 1997. http://www.highbeam.com/doc/1G1-19545620.html.

Atlanta Journal-Constitution. "New Cox Unit to Focus on the Internet." July 12, 1996: D1.

Associated Press. "Analyst: Digital Fees Not Enough to Help NY Times." October 26, 2011. http://finance.yahoo.com/news/Analyst-Digital-fees-not-apf-763804156.html.

Atwater, Tony, Carrie Heeter, and Natalie Brown. "Foreshadowing the Electronic Publishing Age: First Exposures to Viewtron." *Journalism Quarterly* 62, no. 4 (1985): 807–815.

Aumente, Jerome. *New Electronic Pathways: Videotex, Teletext, and Online Databases.* Newbury Park, CA: Sage, 1987.

Bagdikian, Ben H. *The Information Machines: Their Impact on Men and the Media.* New York: Harper & Row, 1971.

———. *The New Media Monopoly.* Boston, MA: Beacon Press, 2004.

Baldwin, Thomas F., D. Stevens McVoy, and Charles Steinfield. *Convergence: Integrating Media, Information and Communication.* Thousand Oaks, CA: Sage, 1996.

Barker, Gerry. "Back to the Future." *Quill,* January 1994, 45–47.

Barrett, Donna, and Randy Siegel, "Time for Newspaper Folks to Fight Back! Here's How." *Editor & Publisher,* January 14, 2009. http://www.editorandpublisher.com/Columns/Article/Time-for-Newspaper-Folks-to-Fight-Back-Here-s-How.

Battelle, John. "The Wizard of Ads." *Business 2.0,* October 2005, 119–121.

Becker, Lee B., Sharon Dunwoody, and Sheizaf Rafaeli. "Cable's Impact on Use of Other News Media." *Journal of Broadcasting* 27, no. 2 (Spring 1983): 127–140.

Bender, Walter, Pascal Chesnais, Sara Elo, Alan Shaw, and Michelle Shaw. "Enriching Communities: Harbingers of News in the Future." *IBM Systems Journal* 35, no. 3&4 (1996): 369–380.

Beniger, James R. *The Control Revolution: Technological and Economic Origins of the Information Society.* Cambridge, MA: Harvard University Press, 1986.

Besen, Stanley M., and Robert W. Crandall. "The Deregulation of Cable Television." *Law and Contemporary Problems* 44, no. 1 (1981): 77–119.

Boczkowski, Pablo J. *Digitizing the News: Innovation in Online Newspapers.* Cambridge, MA: MIT Press, 2004.

———. "The Mutual Shaping of Technology and Society in Videotex Newspapers: Beyond the Diffusion and Social Shaping Perspectives." *Information Society* 20, no. 4 (2004): 255–267.

Bogart, Leo. "Expect No Substitutes: Printed Newspapers Will Ride Out the Electronic Wave." *Presstime,* April 1996, 41.

———. "There's Plenty To Do Before the Unthinkable Happens." *Presstime,* March 1992, 58.

Bontis, Nick, and Jason Mill. "Dot-Bomb Post-Mortem: Web-Based Metrics and Internet Stock Prices." *Quarterly Journal of Electronic Commerce* 4, no. 1 (2004): 1–25.

Booker, Ellis. "Consumer Videotex: The Perilous Path (Prospects for the Videotex Industry in the U.S.)." *Telephony,* June 27, 1988. http://business.highbeam.com/137453/article-1G1-6525332/consumer-videotex-perilous-path.

Bordewijk, J. L., and B. Van Kaam. "Towards a New Classification of Tele-Information Services." *InterMedia* 14, no. 1 (1986): 16–21.

Bovee, Warren G. *Discovering Journalism*. Westport, CT: Greenwood Press, 1999.

Bowman, Shayne, and Chris Willis. "The Future Is Here, But Do News Media Companies See It?" *Nieman Reports*, December 22, 2005. http://www.nieman.harvard.edu/reports/article/100558/The-Future-Is-Here-But-Do-News-Media-Companies-See-It.aspx.

Brooke, Collin G. "Cybercommunities and McLuhan: A Retrospect." In *Rhetoric, the Polls and the Global Village*, edited by C. Jan Swearingen and Dave Pruett, 23–26. Malwah, NJ: Lawrence Erlbaum Associates, 1999.

Brown, Charles L. "Report of the Chairman of the Board." *American Telephone & Telegraph Company 1980 Annual Report*. February 6, 1981, 2–3. http://www.porticus.org/bell/att/1980/att_1980.htm.

Bryant, Jennings. "Message Features and Entertainment Effects." In *Message Effects in Communication Science*, edited by James J. Bradac, 231–262. Newbury Park, CA: Sage, 1989.

Brynjolfsson, Erik, and Adam Saunders. *Wired for Innovation: How Information Technology Is Reshaping the Economy*. Cambridge, MA: MIT Press, 2010.

Bucy, Erik P. "Interactivity in Society: Locating an Elusive Concept." *The Information Society*, no. 20 (2004): 373–383.

———. "The Debate." *The Information Society,* no. 20 (2004): 371.

Buffett, Warren. "Chairman's Letter." *Berkshire Hathaway Inc. 1984 Annual Report*, February 25, 1985. http://www.berkshirehathaway.com/letters/1984.html.

Burgess, John. "Firms Face Questions of Technology's Timing, Cost." *Washington Post*, February 14, 1993. http://www.highbeam.com /doc/1P2-932680.html.

Burroughs, Elise. "Crisis in Classifieds." *Presstime*, May 1996, 39–42.

Butler, Jacalyn Klein, and Kurt E. M. Kent. "Potential Impact of Videotext on Newspapers." *Newspaper Research Journal* 5, no. 1 (1983): 3–12.

Cameron, Glen T., Patricia A. Curtin, Barry Hollander, Glen Nowak, and Scott A. Shamp. "Electronic Newspapers: Toward a Research Agenda." *Journal of Mediated Communication* 11, no. 1 (1996): 3–53.

Carlson, David. "Online Timeline." *Nieman Reports*, Fall 2005, 45–83.

———. "The News Media's 30-Year Hibernation: Online Newspapers Are Not Creative. They Are Not Interactive. They're Too Much Like Newspapers." *Nieman Reports,* Fall 2005, 68–71.

Carr, David F. "Searching for TV-Newspaper Convergence: Studies Show Little Cooperation in Station-Newspaper Partnerships." *Broadcasting & Cable*, October 2008. http://www.broadcastingcable.com/article/95291-Searching_for_TV_Newspaper_Convergence.php.

Carter, Margaret G. "Online Advertising." *Presstime*, March 1995, 38–40.

Caruso, Denise. "Show Me the Money!" *Columbia Journalism Review*, July/August 1997. http://www.caruso.com/work/index-additional-publications/show-me-the-money/.

Chaffin, Joshua. "Knight-Ridder Investor Pushes for Company Sale." *Financial Times*, November 1, 2005. http://www.ft.com/cms/s/0/df17f320-4b2d-11da-aadc-0000779e2340.html.

Chan-Olmsted, Sylvia M. "Mergers, Acquisitions, and Convergence: The Strategic Alliances of Broadcasting, Cable Television, and Telephone Services." *Journal of Media Economics* 11, no. 3 (1998): 33–46.

Chanin, Abe S. "Officials of England's 'Electronic Newspaper' Predict No Early Demise of Print Media." *Presstime*, December 1980, 6–7.

Chon, Bum S., Junho Choi, George A. Barnett, James A. Danowski, and Sung-Hee Joo. "A Structural Analysis of Media Convergence: Cross-Industry Mergers and Acquisitions in the Information Industries." *Journal of Media Economics* 16, no. 3 (2003): 141–157.

Chyi, Hsiang I., and George Sylvie. "Competing with Whom? Where? And How? A Structural Analysis of the Electronic Newspaper Market." *Journal of Media Economics* 11, no. 2 (1998): 1–18.

———. "Online Newspapers in the U.S.: Perceptions of Markets, Products, Revenue and Competition." *International Journal of Media Management* 2, no. 2 (Summer 2000): 69–77.

Cole, Barry G. *After the Breakup: Assessing the New Post-AT&T Divestiture Era.* New York: Columbia University Press, 1991.

Cole, David M. "Industry Saddened, Chastened by Demise of NCN." *NewsInc.*, March 16, 1998. http://www.newsinc.net/morgue/1998/980316/ncn.HTML.

———. "NAA's Ad Numbers May Be Signaling Beginning of the End." *NewsInc.*, March 15, 1999. http://www.newsinc.net/990315sa.html.

———. "Planner: The Virtual Library; Yesterday's Paper Will Provide Tomorrow's Revenue through Digital Access." *Presstime*, September 1995, 47–53.

———. "Webservations." *Presstime*, July/August 1995, 69.

Colon, Aly. "The Multimedia Newsroom." *Columbia Journalism Review*, May/June 2000. http://www.mediageneral.com/newscenter/cjr1.htm.

Compaine, Benjamin M. "The '80s: An Overview." *Presstime*, January 1980, 6–8.

———. *The Newspaper Industry in the 1980s: An Assessment of Economics and Technology.* White Plains, NY: Knowledge Industry Publications, 1980.

Computer Industry Almanac. "25-Year PC Anniversary Statistics." August 14, 2006. http://www.c-i-a.com/pr0806.htm.

———. "The U.S. Now Has One Computer for Every Three People." April 28, 1995. http://www.c-i-a.com/pr0495.htm.

Conhaim, Wallys W. "Developing Videotex as a Consumer Medium." *Information Today*, March 1, 1990. http://www.highbeam.com/doc/1G1-8865293.html.

———. "Newspapers in Transition, Part 2: Envisioning the Future of Newspapers." *Information Today*, October 2006. http://www.highbeam.com/doc/1G1-153051780.html.

Cox Enterprises Inc. *CIMCities Unit Option Plan*. Atlanta, GA: Human Resources, 1999.

Crichton, Michael. "Mediasaurus." Wired, September/October 1993. http://www.wired. com/wired/archive/1.04/mediasaurus.html?topic=&topic_set=.

Criner, Kathleen. "Newspapers at Mid-Decade and Beyond: Telecommunications." *Presstime*, January 1985, 26.

Criner, Kathleen, and Jane Wilson. "An Uncertain Marketplace Takes Telecommunications Toll." *Presstime*, December 1984, 23–25.

———."Telecommunications History Is Short but Stormy." *Presstime*, October 1984, 24–26.

Criner, Kathleen, and Raymond Gallagher. "Newspaper-Cable TV Services: Current Activities in Channel Leasing and Other Local Service Ventures." *Presstime*, March 1982, A3–A8.

Crowe, Thomas K., and Michael J. Jones. "Countdown to N11." *Presstime*, September 1993, 33.

Cuadra, Carlos A. "A Brief Introduction to Electronic Publishing." *Electronic Publishing Review* 1, no. 1 (1981): 29–34.

Cunningham, Brent. "Rush to the Web—and Away from It: New Media Were All So Exciting and Breathtaking. And Then Came the Crash." *Columbia Journalism Review*, November 2001, 120–121.

Damsgaard, Jan. "Managing an Internet Portal." *Communications of the Association for Information Systems* 9 (2002): 408–420.

Davis, Nancy M. "Profiles: Frank A. Daniels III; George Schlukbier." *Presstime*, November 1994, 24–25.

Dominick, Joseph R. *The Dynamics of Mass Communication*. New York: McGraw-Hill, 2009.

Dotinga, Randy. "An Urge to Converge." *MediaWeek*, May 12, 2003, 9.

Dowling, Michael, Christian Lechner, and Bodo Thielmann. "Convergence—Innovation and Change of Market Structures between Television and Online Services." *Electronic Markets Journal* 8, no. 4 (1998): 31–35.

Downes, Edward J., and Sally J. McMillan. "Defining Interactivity: A Qualitative Identification of Key Dimension." *New Media & Society* 2, no. 2 (2000): 157–179.

Dugan, I. Jeanne. "New-Media Meltdown at New Century: How a Big Online Newspaper Venture Bit the Dust." *BusinessWeek*, March 12, 1998. http://www.businessweek. com/1998/12/b3570103.htm.

duPlessis, Renee, and Xigen Li. "Cross-Media Partnership and Its Effect on Technological Convergence of Online News Content: A Content Analysis of 100 Internet Newspapers." In *Internet Newspapers: The Making of a Mainstream Medium*, edited by Xigen Li, 159–175. Mahwah, NJ: Lawrence Erlbaum Associates, 2006.

Economist. "Another Brick in the Wall: The Rapid Rise of Newspaper Paywalls." October 8, 2011. http://www.economist.com/node/21531479.

Editor & Publisher. "It's a Gee Whiz World." February 1, 1986, 8.

Edmonds, Rick. "Newspapers: Economics," *The State of the News Media: An Annual Report on American Journalism 2005*. http://stateofthemedia.org/2005/newspapers-intro/economics/.

Effros, Stephen R. "The Reality Behind the Cable TV Boom." *Presstime*, December 1982, 25.

Elder, Sean. "The Return of the Dead-Tree Media." *Salon.com*, March 15, 2000, http://www.salon.com/media/col/elde/2000/03/15/merger.

eMarketer. "Mobile Passes Print in Time-Spent Among US Adults." December 12, 2011, press release. http://www.emarketer.com/PressRelease.aspx?R=1008732.

Emery, Edwin, and Michael Emery. *The Press and America: An Interpretive History of the Mass Media*. 4th ed. Englewood Cliffs, NJ: Prentice-Hall, 1978.

Erburu, Robert F. "We Can Win Information Services Fight Because Newspaper Position Is Right." *Presstime*, January 1992, 5.

Fancher, Mike. "Disturbing Findings from New Survey." *Seattle Times*, July 31, 2005. http://seattletimes.nwsource.com/html/localnews/2002412302_fancher31.html.

Finberg, Howard I. "Before the Web, There Was Viewtron." *Poynter Online*, October 27, 2003. http://www.poynter.org/uncategorized/17738/before-the-web-there-was-viewtron/.

———. "Viewtron Remembered Roundtable." *Poynter Online*, October 29, 2003. http://www.poynter.org/uncategorized/17740/viewtron-remembered-roundtable/.

Fine, Jon. "The Daily Paper of Tomorrow: It Won't Look the Same. But with Reimagining, the Local Daily Ain't Dead Yet." *BusinessWeek Online*, January 9, 2006. http://www.businessweek.com/print/magazine/content/06_02/b3966026.htm?chan=gl.

Fink, Conrad C. *Strategic Newspaper Management*. Needham Heights, MA: Allyn & Bacon, 1996.

Finkle, Jim. "Cox Interactive's Orange County, Calif., Web Portal to Shut Down." *Orange County Register*, November 22, 2002. http://www.highbeam.com/doc/1G1-94942245.html.

Finnegan, John R., and Kasisomayjula Viswanath. "Community Ties and Use of Cable TV and Newspapers in a Midwest Suburb." *Journalism Quarterly* 65, no. 2 (1988): 456–463, 473.

Flanigan, James. "Newspapers Aren't Dying, They're Just Turning a Page." *Los Angeles Times*, March 19, 2000, C1.

Flichy, Patrice. "The Construction of New Digital Media." *New Media & Society* 1, no. 1 (1999): 33–39.

Folio: The Magazine for Magazine Management. "Advertising Agencies Link Ads, Editorial in Electronic Publications." April 1, 1984. http://www.highbeam.com/doc/1G1-3206743.html.

Fost, Dan. "Rumors of Web's Demise Greatly Exaggerated: Old Media's Retreat from Internet Spending Doesn't Signify Surrender." *SFGate.com*, January 11, 2001.

http://articles.sfgate.com/2001-01-11/business/17578094_1_industry-standard-cable-television-new-york-times-digital.

Friedman, Barbara. "Cable TV: Now Seems to Be a Good Time to Sell or Consolidate System, and Newspapers Are Reaping Benefits Both Ways." *Presstime*, February 1986, 5–7.

———. "Videotex Languor Claims Victims." *Presstime*, April 1986, 56.

———. "Waivers: Bell Regionals Get Green Light to Expand into New Areas." *Presstime*, July 1985, 14–16.

Fuller, Jack. "Merging Media to Create an Interactive Market: New Strategies Are Used to Fund the Expensive Business of Newsgathering." *Nieman Reports*, Winter 2000, 33–35.

Fulton, Kate. "A Tour of Our Uncertain Future." *Columbia Journalism Review*, March/April 1996, 19–26.

Genovese, Margaret. "All Eyes Are on Viewtron Screen Test." *Presstime*, December 1983, 22–23.

———. "Electronics Boom a Gamble or Sure Bet?" *Presstime*, December 1980, 4–7.

———. "Latest Tack for Papers in Cable TV Is to Video." *Presstime,* October 1983, 20–22.

———. "Newspapers Channel Interest in Cable TV." *Presstime*, March 1980, 4–7.

———. "Newspapers Keep Marching to Videotex." *Presstime*, July 1982, 28–29.

———. "PCs No Longer Seen as a Threat." *Presstime*, June 1984, 20.

———. "Viewtron Partners Disappointed over Lackluster Sales." *Presstime*, June 1984, 40.

———. "'VIS'ion of the Future: 'Voice Information' Offers a Relatively Low-Cost Way to Dabble in New Telecommunications Services." *Presstime*, June 1987, 24–29.

Genovese, Margaret, and C. David Rambo."'BISON' Bites Dust, but 'STAR-Text' Is Launched." *Presstime*, June 1982, 16.

Gershon, Richard A., and Abubakar D. Alhassan. "AOL Time Warner & WorldCom Inc.: Corporate Governance and Diffusion of Authority." Paper presented at the 6th World Media Economics Conference, Montreal, Canada, May 12–15, 2004.

Ghosh, Deb. "Videotex Systems." *Database* 23, no. 3 (1992): 19–26.

Gillen, Albert J. "Letter from the President: The Waiting Is Over; Touch the Future." *Viewtron Magazine*, 1983, 2.

Gillette, Becky. "Newspapers, Television Finding Ways to Make Websites Pay." *Mississippi Business Journal*, September 8, 2003. http://msbusiness.com/2003/09/newspapers-television-finding-ways-to-make-web-sites-pay/.

Gipson, Melinda. "@ Your Name Here." *Presstime*, November 1995, 11.

———. "Behold the Megawebsite." *Presstime*, December 1995, 11.

———. "Bill's On-Line AdVenture." *Presstime*, October 1995, 28–33.

———. "New Media: Newspapers and Related Organizations Continue to Launch On-Line Services." *Presstime*, June 1995, 13.

———. "New Media: Strategic Moves." *Presstime*, July/August 1995, 12.

———. "Playing the Online Game to Win." *Presstime*, July/August 2002, 26–31.

———. "Wiring the Waltons." *Presstime*, May 1995, 14–17.

Gitelman, Lisa. *Always Already New: Media, History and the Data of Culture.* Cambridge, MA: MIT Press, 2006.

Glaberson, William. "The Media Business: In San Jose, Knight-Ridder Tests a Newspaper Frontier." *New York Times*, February 7, 1994. http://www.nytimes.com/1994/02/07/business/the-media-business-in-san-jose-knight-ridder-tests-a-newspaper-frontier.html?pagewanted=all&src=pm.

Glaser, Mark. "Annals of Integration: New York Times and Times Digital." *MediaDailyNews*, November 7, 2005. http://www.mediapost.com/publications/article/35985/annals-of-integration-new-york-times-and-times-di.html.

Goldstein, Bill. "A Few Veterans Look Back: Considering the Early Years of Internet Journalism." *New York Times*, January 22, 2001. http://partners.nytimes.com/library/tech/reference/roundtable.html.

Gomery, Douglas. "Media Economics: Terms of Analysis." *Critical Studies in Mass Communication* 6 (1989): 43–60.

Google. "Google Announces Record Revenues for Fourth Quarter and Fiscal Year 2004." February 1, 2005, press release. http://investor.google.com/earnings/2004/Q4_google_earnings.html.

Gordon, Rich. "The Meanings and Implications of Convergence." In *Digital Journalism: Emerging Media and the Changing Horizons of Journalism*, edited by Kevin Kawamoto, 57–73. Lanham, MD: Rowman & Littlefield, 2003.

Green, Nicola, and Leslie Haddon. *Mobile Communications: An Introduction to New Media.* New York: Berg, 2009.

Groves, Miles E. "1996 Forecast: A Look at What Was, Why and What May Be." *Presstime*, January 1996, 24–26.

Gunther, Marc. "Publish or Perish?" *Fortune*, January 10, 2000, http://money.cnn.com/magazines/fortune/fortune_archive/2000/01/10/271744/.

Gurnsey, John. *Electronic Document Delivery—III: Electronic Publishing Trends in the United States and Europe.* Medford, NJ: Learned Information, 1982.

Hartley, John. *Communication, Cultural and Media Studies: The Key Concepts.* 3rd ed. New York: Routledge, 2002.

Hecht, Jeff. "Information Services Search for Identity." *High Technology*, no. 3 (1983): 58–65.

Henke, Lucy L., and Thomas R. Donahue. "Teletext Viewing Habits and Preferences." *Journalism Quarterly* 63, no. 3 (1986): 542–553.

Hickey, Neil. "Coping with Mega-Mergers." *Columbia Journalism Review*, March/April 2000, 16–20.

Hilder, David B. "Will Public Like Newspapers on TV?" *Atlanta Journal and Constitution*, June 8, 1980, 16B.

Hill, Miriam. "Sale Would End Family's Empire After 114 Years: P. Anthony Ridder, Pushed by Investors to Take Bids, Would Be Out as CEO. The Ridders Bought Their First Newspaper in 1892." *Philadelphia Inquirer*, March 13, 2006. http://articles.philly.com/2006-03-13/news/25415023_1_tony-ridder-knight-ridder-private-capital-management.

Hinton, Sam M. "Portal Sites: Emerging Structures for Internet Control." *Research Report No. 6*. La Trobe University Online Media Program, January 1999, 1–43.

Hodge, Bob. "How the Medium Is the Message in the Unconscious of 'America Online.'" *Visual Communication* 2, no. 3 (2003): 341–353.

Hollander, Barry. "Talk Radio, Videotext and the Information Superhighway." *Editor & Publisher*, July 16, 1994, 60–61.

Holm, Erik, and Christopher S. Stewart. "In Deal, Buffett Departs from Type." *Wall Street Journal*, December 1, 2011, C1–2.

Holt, Jennifer, and Alisa Perren, eds. *Media Industries: History, Theory, and Method*. Malden, MA: Blackwell-Wiley, 2009.

Hooker, Michael. "A Library on My Belt." In *Come the Millennium: Interviews on the Shape of Our Future*, American Society of Newspaper Editors, 45–49. Kansas City, MO: Andrews and McMeel, 1994.

Ingle, Robert. "Newspaper vs. On-Line Versions: A Discussion of the Old and New Media." *Nieman Reports*, June 22, 1995. http://www.highbeam.com/doc/1G1-17260055.html.

Ives, Nat. "What Matters Most in Magazine and Newspaper iPad Apps? Quality, Report Says." *Advertising Age*, October 27, 2011. http://adage.com/article/mediaworks/quality-matters-magazine-newspaper-ipad-apps/230673/.

Jensen, Jens F. "Interactivity: Tracing a New Concept in Media and Communication Studies." *Nordicom Review* 19, no. 1 (1998): 185–204.

Johnson, Robert M. "Use It or Lose It: Electronic Publishing; Between Now and 1990, Newspapers Must Act to Convince the Court They Are Serious About Offering High-Tech Information Services." *Presstime*, February 1988, 10–12.

Jones, Steven. "The Bias of the Web." In *The World Wide Web and Contemporary Cultural Theory*, edited by Andrew Herman and Thomas Swiss, 171–182. New York: Routledge, 2000.

Jones, Tim, and Gary Marx. "Weaned on Crisis, Landing on Top: The Man Behind AOL Is an Online Visionary Who's Had His Share of Crashes." *Chicago Tribune*, January 16, 2000. http://articles.chicagotribune.com/2000-01-16/news/0001160155_1_aol-time-warner-steve-case-aol-plans.

Jung, Jaemin. "The Bigger, the Better? Measuring the Financial Health of Media Firms." *International Journal on Media Management* 5, no. 4 (Winter 2003): 237–250.

Kahaner, Larry. "'Baby' Bells Grow into Giants: Newspapers Fear RBOC's May Become Information Providers over Their Own Lines." *Presstime*, February 1990, 18–27.

Kaletsky, Anatole. *Capitalism 4.0: The Birth of a New Economy in the Aftermath of Crisis.* New York: PublicAffairs, 2010.

Kapor, Mitchell. "Where Is the Digital Highway Really Heading? The Case for a Jeffersonian Information Policy." *Wired*, July/August 1993. http://www.wired.com/wired/archive/1.03/kapor.on.nii.html

Kasler, Dale. "McClatchy Purchase Heralds New Era: CEO Pruitt Says Buying Knight-Ridder Is the Right Move Despite High Stakes." *The Sacramento Bee*, March 14, 2006. http://mediaworkers.org/index.php?ID=1840.

Katz, Jon. "Online or Not, Newspapers Suck." *Wired*, September 1994, http://www.wired.com/wired/archive/2.09/news.suck_pr.html.

Kauffman, Jack. "Newspapers See Opportunities in Electronic Classified Ads." *Presstime*, October 1980, 35.

Ketterer, Stan, Tom Weir, Steven J. Smethers, and James Back. "Case Study Shows Limited Benefits of Convergence." *Newspaper Research Journal* 25, no. 3 (Summer 2004): 52–65.

Kilker, Julian A. "Shaping Convergence Media." *Convergence* 9, no. 3 (2003): 20–39.

Kinsley, Michael. "America Is Losing the Information Revolution." *Washington Post*, January 8, 1987. http://www.highbeam.com./5554/article-1P2-1299682/america-losing-information-revolution-buzz-buzz.

———. "The Latest Moos: Newspaper Publishers Protect Their Cash Cows." *New Republic: A Weekly Journal of Opinion*, September 16, 1981, 20–22.

Kiousis, Spiro. "Interactivity: A Concept Explication." *New Media & Society* 4, no. 3 (2002): 355–383.

Kirchhoff, Suzanne M. "The U.S. Newspaper Industry in Transition." *Congressional Research Service: Report for Congress*, July 8, 2009, 1–23.

Kist, Joost. *Electronic Publishing: Looking for a Blueprint.* New York: Croom Helm, 1987.

Knight-Ridder Inc. "Knight-Ridder Closes Information Design Laboratory." August 1, 1995, press release. http://www.highbeam.com/doc/1G1-17126599.html.

———. "Knight-Ridder Inc. Agrees to Purchase Dialog Information Services, Inc. from Lockheed Corp. for $353 Million." July 11, 1988, press release. http://www.highbeam.com/doc/1G1-6821481.html.

———. "Knight-Ridder Announces Exploration of Strategic Alternatives." November 14, 2005, press release. http://www.highbeam.com/doc/1G1-138646176.html.

Knowledge@Wharton. "All the News That's Fit to . . . Aggregate, Download, Blog: Are Newspapers Yesterday's News?," March 22, 2006. http://knowledge.wharton.upenn.edu/article.cfm?articleid=1425.

Kolodzy, Janet. "Everything That Rises." *Columbia Journalism Review*, July/August 2003, 61.

Koziol, Michael. "Remix the Media Mix." *BtoB*, April 3, 2006. http://www.highbeam.com/doc/1G1-144218241.html

Kyrish, Sandy. "From Videotext to the Internet: Lessons from Online Services 1981–1996." *Research Report No. 1,* La Trobe University Online Media Program, August 1996, 1–31.

———. "Lessons from a 'Predictive History': What Videotex Told Us about the World Wide Web." *Convergence: The International Journal of Research into New Media Technologies* 7, no. 4 (2001): 10–29.

Laakaniemi, Ray. "The Computer Connection: America's First Computer-Delivered Newspaper." *Newspaper Research Journal* 2, no. 4 (1981): 61–68.

Lacy, Stephen, and Hugh J. Martin. "Competition, Circulation and Advertising." *Newspaper Research Journal* 25, no. 1 (Winter 2004): 18–39.

Lancaster, F.W. *Toward Paperless Information Systems.* New York: Academic Press, 1978.

Lapham, Chris. "The Evolution of the Newspaper of the Future." *CMC Magazine,* July 1, 1995. http://www.ibiblio.org/cmc/mag/1995/jul/lapham.html.

Lawson-Borders, Gracie. "Integrating New Media and Old Media: Seven Observations of Convergence as a Strategy for Best Practices in Media Organizations." *International Journal on Media Management* 5, no. 2 (Summer 2003): 91–99.

Leavitt, Harold, and Thomas Whisler. "Management in the 1980s." *Harvard Business Review* 36, no. 6 (1958): 41–48.

Leckner, Sara. "Is the Medium the Message?: The Impact of Digital Media on the Newspaper Concept." PhD diss., Royal Institute of Technology, 2007.

Lee, William Chee-Leong. "Clash of the Titans: Impact of Convergence and Divergence on Digital Media." Master's thesis, Massachusetts Institute of Technology, 2003.

Lee, Ya-Ching. "Newspaper Online Services: A Successful Business?; Lessons Learned from Videotext Failure." Department of Telecommunications, Indiana University, July 1999. http://130.203.133.150/viewdoc/summary?doi=10.1.1.42.9681.

LeGates, John C. "Many Noses Being Poked Under Information Tent." *Presstime,* February 1981: 23–25.

Lehman-Wilzig, Sam, and Nava Cohen-Avigdor. "The Natural Life Cycle of New Media Evolution." *New Media & Society* 6, no. 6 (2006): 707–730.

Lerner, Rita G. "Electronic Publishing." In *New Trends in Documentation and Information: Proceedings of the 39th FID Congress, University of Edinburgh, September 25–28 1978,* edited by Peter J. Taylor, 111–116. London: Aslib, 1980.

Lewis, Peter H. "The Media Business: Big Newspapers to Help Locals on Internet." *New York Times,* April 20, 1995. http://www.nytimes.com/1995/04/20/business/the-media-business-big-newspapers-to-help-locals-on-internet.html?pagewanted=all&src=pm.

Liedtke, Michael. "Knight Ridder's Demise Reflects Sobering Times for U.S. Newspapers." *Associated Press,* June 23, 2006. http://www.highbeam.com/doc/1P1-125713972.html.

———. "Web Growth, Innovation Threatens Papers." *AP Online,* April 15, 2005. http://www.highbeam.com/doc/1P1-107483324.html.

Light, Jennifer S. "Facsimile: A Forgotten 'New Medium' from the 20th Century." *New Media & Society* 8, no. 3 (2006): 355–378.

Lind, Jonas. "Convergence: History of Term Usage and Lessons for Firm Strategists." Paper presented at the 15th Biennial ITS Conference, Berlin, Germany, June 2004.

Littlewood, Thomas B. "A View from 88 Years Ago of Newspapers in Year 2000." *Presstime*, July 1991, 62.

Maddox, Kate. "Joint Report Reveals Online Ad Recovery; Traditional Advertisers Embrace Rich Media, but Not All Sectors Benefitting." *BtoB*, March 8, 2004, http://www.highbeam.com/doc/1G1-114166980.html.

Maguire, W. Terry. "How the Diversity Principle Became the Law of the Land." *Presstime*, October 1982, 22–26.

Maney, Kevin. *Megamedia Shakeout: The Inside Story of the Leaders and Losers in the Exploding Communications Industry*. New York: John Wiley & Sons, 1995.

Mann, Jennifer. "Deal Fuels Debate on Future of Print." *Kansas City Star*, March 14, 2006, D4.

Manovich, Lev. *The Language of New Media*. Cambridge, MA: MIT Press, 2001.

Mansell, Robin. "New Media Competition and Access: The Scarcity-Abundance Dialectic." *New Media & Society* 1, no. 2 (1999): 155–182.

Marbut, Robert G. "Confronting the Telecommunications Issue." *Presstime*, June 1981, 29–32.

Margolies, Dan. "Chain's Sale Is Milestone for Industry, Star." *Kansas City Star*, June 27, 2006, A1.

Markoff, John. "Business Technology: A Free and Simple Computer Link." *New York Times*, December 8, 1993. http://www.nytimes.com/1993/12/08/business/business-technology-a-free-and-simple-computer-link.html?pagewanted=all&src=pm.

———. "17 Companies in Electronic News Venture." *New York Times*, May 7, 1993. http://www.nytimes.com/1993/05/07/business/17-companies-in-electronic-news-venture.html.

Massachusetts Institute of Technology. "Consortium to Probe News Presentation." May 12, 1993, press release. http://web.mit.edu/press/1993/news-0512.html.

Maxwell, Ann, and Wayne Wanta. "Advertising Agencies Reduce Reliance on Newspaper Ads." *Newspaper Research Journal* 22, no. 2 (Spring 2001). http://www.highbeam.com/doc/1G1-79573837.html.

Mayer, William G. "The Polls—Poll Trends: The Rise of the New Media." *Public Opinion Quarterly* 58, no. 1 (1994): 124–146.

McCombs, Maxwell E. "Mass Media in the Marketplace." *Journalism Monographs*, no. 24 (August 1972): 1–104.

McCombs, Maxwell E., and Chaim H. Eyal. "Spending on Mass Media." *Journal of Communication*, no. 30 (Winter 1980): 153–158.

McKenna, Kate. "The Future Is Now: Newspapers Are Overcoming Their Fears of Technology and Launching a Wide Array of Electronic Products." *American Journalism Review*, October 1993. http://www.ajr.org/article.asp?id=1452.

———. "Where the Action Is." *American Journalism Review*, October 1993. http://www. ajr.org/article.asp?id=1454.

McLuhan, Marshall. *Understanding Media: The Extensions of Man.* New York: McGraw-Hill, 1964.

McMillan, Sally J. "Who Pays for Content? Funding in Interactive Media." *Journal of Computer-Mediated Communication*, September 1998. http://www.blackwellsynergy.com/doi/full/10.1111/j.1083- 6101.1998.tb00090.x.

Meisel, J. B., and T. S. Sullivan. "Portals: The New Media Companies." *Journal of Policy, Regulation and Strategy for Telecommunications* 2, no. 5 (2000): 477–486.

Mencke, Chuck. "Farewell to a Good Friend." *Cruisin' with StarText*, March 14, 1997. http://www.webconnection.org/archive/crusback/1997/class0314.htm.

Microsoft Inc. "Microsoft Announces Availability of the Microsoft Network; Consumers Worldwide Invited to Get Online with Windows 95 and MSN." August 24, 1995, press release. http://www.highbeam.com/doc/1G1-17231563.html

Miles, Stephanie. "New Century Networks Regroups." *CNET News.com,* January 9, 1998. http://news.cnet.com/New-Century-Networks-regroups/2100-1001_3-206952. html.

———. "News Megasite Cuts Content, Staff." *CNET News.com*, February 25, 1998. http:// news.cnet.com/News-megasite-cuts-content,-staff/2100-1001_3-208500.html.

Miller, Toby. "The Economy's New Clothes?" *Television & New Media* 1, no. 4 (2000): 371–373.

Millman, Nancy. "Videotext Age Comes to an End: Viewtron Folds; Knight-Ridder Pulls Plug on Unit." *Chicago Sun-Times,* March 18, 1986. http://www.highbeam.com/ doc/1P2-3756791.html.

Mings, Susan M., and Peter B. White. "Profiting from Online News: The Search for Viable Business Models." In *Internet Publishing and Beyond: The Economics of Digital Information and Intellectual Property*, edited by Brian Kahin and Hal R. Varian, 62–96. Cambridge, MA: MIT Press, 2000.

Moeller, Philip. "The High-Tech Trib." *American Journalism Review*, April 1994. http:// www.ajr.org/article.asp?id=1465.

Montagne, Renee. "Part Two of the Series on Media Convergence Looks at a Media Empire in Lawrence, Kansas." Morning Edition produced by National Public Radio, April 14, 2005. http://www.highbeam.com/doc/1P1-107424184.html.

Morton, John. "The Emergence of Convergence." *American Journalism Review*, January/ February 2000, 88.

———. "Papers Will Survive Newest Technology." *American Journalism Review*, June 1993, 48.

———. "When Newspapers Eat Their Seed Corn." *American Journalism Review*, November 1995, 52.

Moss, Mitchell L. "Cable Television: A Technology for Citizens." *University of Detroit Journal of Urban Law* 45, no. 3 (1978): 699–720.

Mowshowitz, Abbe. "Scholarship and Policy Making: The Case of Computer Communications." *Computers and Society* 13, no. 2 (1983): 2–9.

Naisbitt, John. *Megatrends: Ten New Directions Transforming Our Lives.* New York: Warner Books, 1982.

Nerone, John, and Kevin G. Barnhurst. "Beyond Modernism: Digital Design, Americanization and the Future of Newspaper Form." *New Media & Society* 3, no. 4 (2001): 467–482.

Neuharth, Allen H. "Opportunities for Newspapers Will Abound in Exciting '80s." *Presstime,* January 1980, 2.

Neustadt, Richard M. *The Birth of Electronic Publishing: Legal and Economic Issues in Telephone, Cable and Over-the-Air Teletext and Videotext.* White Plains, NY: Knowledge Industry Publications, 1982.

Newspaper Association of America (formerly the Newspaper Publishers Association of America). "Trends and Numbers: Advertising Expenditures." http://www.naa.org/Trends-and-Numbers/Advertising-Expenditures/Annual-All-Categories.aspx.

———. "Trends and Numbers: Circulation." http://www.naa.org/Trends-and-Numbers/Circulation/Newspaper-Circulation-Volume.aspx.

———. "Public Policy: Legal Affairs/Business Operations." http://www.naa.org/Public-Policy/Legal-Affairs/Business-Operations.aspx.

Newspaper Controller. "A Look at Electronic Publishing." November 1982, 6–7.

———. "The Economics of Telecommunications." November 1983, 14–16.

New York Times. "Information Technology: The Times Is Joining New On-Line Service." May 8, 1995. http://www.nytimes.com/1995/05/08/business/information-technology-the-times-is-joining-new-on-line-service.html.

New York Times Co. "Newspaper Establishes Internet Presence." January 22, 1996, press release. http://partners.nytimes.com/library/tech/reference/012296press_release.html.

Noam, Eli M. "Towards an Integrated Communications Market: Overcoming the Local Monopoly of Cable Television." *Federal Communications Law Journal* 34, no. 2 (1982): 209–257.

O'Leary, Mick. "NewsWorks Brings New Depth to Web News." *Information Today,* December 1, 1997. http://www.highbeam.com/doc/1G1-20051760.html.

Osder, Elizabeth Anne. "The Little Browser That Could and the New Media Revolution." *Poynter Online,* February 23, 2004. http://www.poynter.org/uncategorized/21162/the-little-browser-that-could-and-the-new-media-revolution/.

Overholser, Geneva. "Peril, Paranoia and Promise." *Presstime,* April 1996, 54.

Palser, Barb. "The Ins and Outs of iPad Apps: Here's How Top News Outlets Are Presenting Material on the Popular Tablet Device." *American Journalism Review,* March 2011. http://www.ajr.org/article.asp?id=5027.

Patten, David A. *Newspapers and New Media.* White Plains, NY: Knowledge Industry Publications, 1986.

Patterson, Scott. "Buffett Sees 'Unending Losses' for Many Newspapers." *WSJ.com*, May 2, 2009. http://blogs.wsj.com/marketbeat/2009/05/02/ buffett-sees- unending-losses-for-many-newspapers/.

Pavlik, John. "The Future of Online Journalism: Bonanza or Black Hole?" *Columbia Journalism Review*, July/August 1997, 30–36.

———. "The Impact of Technology on Journalism." *Journalism Studies* 1, no. 2 (May 2000): 229–237.

Pelline, Jeff. "Cox Muscles into Local Markets." *CNET News.com*, February 20, 1997. http://news.cnet.com/Cox-muscles-into-local-markets/2100-1023_3-272491.html.

———. "Print to Try Packaging with NewsWorks." *CNET News.com*, June 19, 1997. http:// news.cnet.com/Print-to-try-packaging-with-NewsWorks/2100-1023_3-200427. html.

Peng, Foo Yeuh, Naphtali Irene Tham, and Hao Xiaoming. "Trends in Online Newspapers: A Look at the U.S. Web." *Newspaper Research Journal* 20, no. 2 (March 22, 1999). http://www.highbeam.com/doc/1G1-55070892.html.

Peskin, Dale. "Slaying 'The Mediasaurus:' An Editor Debunks Michael Crichton's View of Media Obsolescence." *Presstime*, June 1994, 51–52.

Peters, Benjamin. "And Lead Us Not into Thinking the New Is New: A Bibliographic Case for New Media History." *New Media & Society* 11, no. 1&2 (2009): 13–30.

Peterson, Iver. "New Service Skims 150 Newspapers for Its Users." *New York Times*, June 30, 1997. http://www.nytimes.com/1997/06/30/business/new-service-skims-150- newspapers-for-its-users.html.

Picard, Robert G. "Capital Crisis in the Profitable Newspaper Industry." *Nieman Reports,* Winter 2006, 10–12.

———. "Changing Business Models of Online Content Services: Their Implications for Multimedia and Other Content Producers." *International Journal on Media Management* 2, no. 2 (Summer 2000): 60–68.

Potter, Walt. "Bells' Worries Range Far and Wide." *Presstime*, September 1993, 28.

———. "Broader Fight Looms in House Over Restrictions on RBOCs." *Presstime*, April 1992, 38–39.

———. "Changed Marketplace, Changed Attitudes: Newspapers and the Bells Have Changed Their Tone as They Seek Common Ground." *Presstime*, September 1993, 27–30.

———. "Cox Seeks Power in Partnerships." *Presstime*, August 1993, 19.

———. "The Deja View: Publishers' Roles in the Development of, and Adaptation to, Broadcast Offer Uplift in the New-Media Age." *Presstime*, November 1994, 27–29.

———. "Early Newspaper Efforts as Information Providers on Cable Were Enthusiastic and Imaginative but Fell Short." *Presstime*, March 1991, 10–11.

———. "From Concept to Strategic Plan: Dow Jones Makes Some Connections." *Presstime*, August 1992, 6–7.

———. "A Guide to Online Avenues." *Presstime*, January 1995, S1–S16.

———. "How to Build Your Own Audiotex System." *Presstime*, July 1991, 10–12.

———."Joe Reader Meets the E-Paper: You've Read about Them. We Decided to Actually Read Them." *Presstime*, October 1993, 44–47.

———. "A Local Look for Cable-TV News." *Presstime*, March 1991, 9–11.

———."Media Merger Mania II: The Quest for Content." *Presstime*, January 1994, 27–29.

———."Papers Battle RBOC's on All Fronts." *Presstime,* November 1991, 54–57.

Presstime. "Assessing the Challengers." August 1992, 36.

———. "Association Now at Home on the Web." July/August 1995, 6.

———. "AT&T Petitions Court for Removal of Electronic Publishing Curbs." May 1989, 70.

———. "Atlanta's Custom Connection." April 1994, 8.

———. "BellSouth, ANPA Clash over Progress of Gateway." June 1989, 64.

———. "Cable TV Start-Ups Show Decline since Peak in 1982." December 1983, 24.

———. "CompuServe Evaluation Marks End to Second Year of Operation." June 1982, 17.

———. "Convention Report: A Classified Wake-Up Call." May 1996, 18.

———. "Convention Report: AT&T's Brown Rebuts Press Concerns." May 1982, 21–22.

———. "The Content Kings' Conundrum." July/August 1995, 27.

———. "'Diversity Principle' Reaffirmed." October 1987, 42–43.

———. "Electronic Publishers Describe How They Do It." July 1982, 18.

———. "Future Bound: Products, Systems and Practices Preview What's Next." July/August 1995, 30–31.

———. "Interactivity: On-line's Edge." July/August 1995, 50.

———. "More Papers Entered Cable TV for Competition than Profits." February 1984, 30.

———. "Newspaper Jobs Now Total 432,000; Most in Nation." May 1981, 50.

———. "Newspapers Must Adopt New Technology to Improve Role of 'Print Gateway.'" July 1990, 30.

———. "'No Revenue Potential' Ends Cable Service." August 1983, 12.

———. "Panel's Advice: Get in Gear; Content, Identity Drive Development." May 1994, 27.

———. "Papers Must Deliver Information in Many New Ways, Erburu Says: Traditional Print Publication Is Not for Everyone." July 1991, 26.

———. "Partner Power and the Next Media Age." July/August 1994, 38–39.

———. "Peters: Get Crazy, or Else." July 1993, 30.

———. "Print Still Prime." July/August 1994, 34.

———. "Putting It Together: The Las Vegas Trade Show Proved a Technological Oasis of Shared Ideas, Strategies and Solutions." July/August 1994, 30–34.

————. "Regional Bell Companies Mount Strong Fight to Enter Electronic Publishing." November 1989, 42.

————. "Six Papers Blaze Career Path." November 1995, 15.

————. "Telecommunications Is a 'Pathway to the Future.'" July 1990, 34–35.

————. "Telecommunications Notes." June 1982, 17.

————. "Telecommunications Technology Beckons, But Is It a Boon or a Bane?" May 1989, 46–47.

————. "Three BOCs Rebuffed in Bid to Offer Electronic Yellow Pages." July 1989, 61.

————. "Tomorrow Today." May 1993, 46.

————. "Tomorrow's Technology: Be There Now." July 1993, 40.

————. "Voice-Information Services Provide Successful Newspaper Ventures." May 1990, 43.

Pryor, Larry. "The Videotex Debacle." *American Journalism Review*, November 1994, 40–42.

Rainie, Lee, and Peter Bell. "The Numbers That Count." *New Media & Society* 6, no. 1 (2004): 44–54.

Rambo, David C. "AT&T's Home Video Plans Eyed Warily." *Presstime*, May 1980, 48–49.

————. "Audiotex: Dial-Up Services Now Provide Weather, Sports Scores; Could Bridge the Gap to More Complex Systems." *Presstime*, October 1984, 28–30.

————. "Clouding the Future? Electronic Publishers Face Legal Questions on Variety of Issues." *Presstime*, October 1982, 4–9.

————. "Court Approves AT&T Consent Decree." *Presstime*, September 1982, 10–12.

————. "Divestiture: How Will Breakup of Ma Bell Affect Newspapers?" *Presstime*, January 1984, 10–11.

————. "It's Still a 'Maybe' Market for New Technologies." *Presstime*, January 1983, 20–22.

————. "Legislative Activity Focusing on AT&T Deregulation Bills." *Presstime*, July 1980, 44.

————. "New Services Stir Variety of Questions on Marketing." *Presstime*, October 1981, 24–27.

————. "Newspaper Companies Go Back to Basics." *Presstime*, January 1987, 22–30.

Raymond, Darrell R. "Why Videotex Is (Still) a Failure." *Canadian Journal of Information Science* 14, no. 1 (March 1989): 27–38.

Reed, David. "Cabletext: Is it Newspapering?" *The Bulletin of the American Society of Newspaper Editors*, November 1982, 23.

Reid, Robert H. *Architects of the Web: 1,000 Days That Built the Future of Business*. New York: John Wiley, 1997.

Rheingold, Howard. *The Virtual Community: Homesteading on the Electronic Frontier*. Cambridge, MA: MIT Press, 2000.

Rice, Ronald E. "Artifacts and Paradoxes in New Media," *New Media & Society* 1, no. 1 (1999): 24–32.

Rittenhouse, Robert G. "The Market for Wired City Services." *Computers & Society* 10, no. 2 (1979): 2–13.

Robinson, John. "Television and Leisure Time: A New Scenario." *Journal of Communication* 31, no. 1 (Winter 1981): 120–130.

Romenesko, James. "What the Near Future Holds." *Knight Ridder/Tribune News Service*, April 9, 1997, http://www.highbeam.com/doc/1G1-19284325.html.

Rosen, Jay. "Laying the Newspaper Gently Down to Die." *Pressthink*, March 29, 2005. http://journalism.nyu.edu/pubzone/weblogs/pressthink/2005/03/29/nwsp_dwn.html.

Rosenblatt, Roger. "Machine of the Year: A New World Dawns." *Time*, January 3, 1983, 13.

Rosenfeld, Arnold. "Videotext vs. Newspapers: The Press's $64,000 Ifs." *Bulletin of the American Society of Newspaper Editors*, December/January 1982, 27–28.

Rosenthal, Phil. "Media Column." *Chicago Tribune*, March 14, 2006. http://www.highbeam.com/doc/1G1-145520735.html.

Rothenberg, Randall. "Convergence Likely to Render Restriction of Liquor Ads Futile." *Advertising Age*, February 18, 2002, 14.

Russell, J. Thomas. "Advertising in an Electronic World." *Journal of Advertising* 7, no. 4 (1978): 52.

Schlukbier, George. "Internet: Newspapers' Best Strategy?" *Presstime*, January 1995, S14–S15.

Schonfeld, Erick. "The Flickrization of Yahoo." *Business 2.0*, December 2005, 157–164.

———. "Tuning Up Big Media: A Modest Proposal for Saving Time Warner and the Entire Industry from Themselves." *Business 2.0*, April 2006, 61–63.

Scott, Ben. "A Contemporary History of Digital Journalism." *Television & New Media* 6, no. 1 (February 2005): 89–126.

Scully, Vaughan. "Media Market: The Bells Toil for Them." *NewsInc.*, June 1992. http://www.highbeam.com/doc/1G1-43035731.html.

Seelye, Katharine Q. "What-Ifs of a Media Eclipse." *New York Times*, August 27, 2006. http://www.nytimes.com/2006/08/27/business/yourmoney/27knight.html.

Shapiro, Eben. "The Media Business; New Features Are Planned by Prodigy." *New York Times*, September 6, 1990. http://www.nytimes.com/1990/09/06/business/the-media-business-new-features-are-planned-by-prodigy.html.

Shaw, Rochelle. "Portals: An Introduction." *Gartner Technology Overview*, May 23, 2000, 1–4.

Shaw, Russell. "Videotext, the Sequel." *MediaWeek*, February 27, 1995, 14.

Shedden, David. "New Media Timeline (1980)." *Poynter Online*, October 26, 2007. http://www.poynter.org/uncategorized/28725/new-media-timeline-1980/.

————. "New Media Timeline (1982)." *Poynter Online*, October 26, 2007. http://www.poynter.org/uncategorized/28730/new-media-timeline-1982/.

————. "New Media Timeline (1983)." *Poynter Online*, October 26, 2007. http://www.poynter.org/uncategorized/28733/new-media-timeline-1983/.

————. "New Media Timeline (1985)." *Poynter Online*, October 26, 2007. http://www.poynter.org/uncategorized/28739/new-media-timeline-1985/.

————. "New Media Timeline (1988)." *Poynter Online*, October 26, 2007. http://www.poynter.org/uncategorized/28748/new-media-timeline-1988/.

————. "New Media Timeline (1995)." *Poynter Online*, October 26, 2007. http://www.poynter.org/uncategorized/28772/new-media-timeline-1995/.

————. "New Media Timeline (1996)." *Poynter Online*, October 26, 2007. http://www.poynter.org/uncategorized/28775/new-media-timeline-1996/.

Sigel, Efrem. *The Future of Videotext: Worldwide Prospects for Home/Office Electronic Information Services*. White Plains, NY: Knowledge Industry Publications, 1983.

Silverthorne, Sean. "Read All About It! Newspapers Lose Web War: Q&A with Clark Gilbert." *Harvard Business School Working Knowledge*, January 28, 2002. http://hbswk.hbs.edu/item/2738.html.

Singer, Jane B. "Strange Bedfellows? The Diffusion of Convergence in Four News Organizations." *Journalism Studies* 5, no. 1 (February 2004): 3–18.

Smith, Anthony. *Goodbye, Gutenberg: The Newspaper Revolution of the 1980s*. New York: Oxford University Press, 1981.

————. "Transition to Electronics: From a Bright Past to an Uncertain Future." *Bulletin of the American Society of Newspaper Editors*, December/January 1982, 10–12.

Smolkin, Rachel. "Adapt or Die." *American Journalism Review*, June 2006, 16–23.

Snyder, Beth, and Anne Marie Kerwin. "Publishers Fight for Local Dollars in Online Space: Newspapers Get More Aggressive with Web Sites." *Advertising Age*, November 24, 1997. http://www.highbeam.com/doc/1G1-20012344.html.

Sparks, Glenn G. *Media Effects Research: A Basic Overview*. 3rd ed. Boston, MA: Wadsworth, 2010.

Standera, Oldrich. *The Electronic Era of Publishing: An Overview of Concepts, Technologies and Methods*. New York: Elsevier, 1987.

Steinberg, Brian. "Newspaper Woes Are Black and White." *Wall Street Journal*, December 15, 2004, B3.

Sterling, Ellen. "Digital First: What Does It Mean, And Where Will It Take Us?" *Editor & Publisher*, December 13, 2011. http://www.editorandpublisher.com/Features/Article/Digital-First--What-Does-It-Mean--And-Where-Will-It-Take-Us-.

Stevenson, Richard W. "Videotex Players Seek a Workable Formula." *New York Times*, March 25, 1986. http://www.nytimes.com/1986/03/25/business/videotex-players-seek-a-workable-formula.html?pagewanted=all.

Stöber, Rudolf. "What Media Evolution Is: A Theoretical Approach to the History of New Media." *European Journal of Communications* 19, no. 4 (2004): 483–505.

Stromer-Galley, Jennifer. "Interactivity-as-Product and Interactivity-as-Process." *The Information Society*, no. 20 (2004): 391–394.

Sullivan, Carl. "Cox Shifts Web Strategy to Local Control." *Editor & Publisher*, June 13, 2002. http://www.allbusiness.com/services/business-services- miscellaneous- business/4692885-1.html.

Svaldi, Aldo. "Online Outlets Link Newspapers, Profits; Publisher Tells of Internet Boost." *Denver Post*, October 29, 2003. http://www.highbeam.com/doc/1G1-109518324.html.

Taylor, Cathy. "AOL: Won't Tangle with Web." *MediaWeek,* May 27, 1996, 6–7.

Thelen, Gil. "Tampa's Convergence Lessons." *The American Editor,* July 2000, 7–9.

Thielmann, Bodo, and Michael Dowling. "Convergence and Innovation Strategy for Service Provision in Emerging Web-TV Markets." *International Journal on Media Management* 1, no. 1 (Autumn/Winter 1999): 4–9.

Thompson, David R., and Birgit L. Wassmuth. "Few Newspapers Use Online Classified Interactive Features." *Newspaper Research Journal* 22, no. 4 (Fall 2001): 16–27.

Toffler, Alvin. *Future Shock.* New York: Random House, 1970.

———. *Previews and Premises.* New York: Morrow, 1983.

———. *The Third Wave.* New York: Morrow, 1980.

Tompkins, Al, and Aly Colon. "Tampa's Media Trio." *Broadcasting & Cable*, April 10, 2000, 46–53.

Toner, Mark. "Competing for Cyberturf." *Presstime*, September 1997. http://web.archive.org/web/20000307024003/http://www.naa.org/presstime/9709/connex.html.

———. "Getting on Boards." *Presstime*, May 1995, 47–50.

———. "Standards Bearers Enter the Web." *Presstime*, June 1995, 76.

Toner, Mark, and Melinda Gipson. "New Media: On Line." *Presstime*, November 1995, 10–11.

U.S. Bureau of Labor Statistics. "Employment Projections: 2008–2018." December 10, 2009, press release. http://www.bls.gov/news.release/ecopro.nr0.htm.

———. "Occupational Employment Statistics: NAICS 511110 – Newspaper Publishers," May 2010. http://www.bls.gov/oes/current/naics5_511110.htm.

U.S. Census Bureau. *Statistical Abstract of the United States: 1980.* Washington, DC: Department of Commerce, 1980. http://www2.census.gov/prod2/statcomp/documents/1980-01.pdf.

———. *Statistical Abstract of the United States: 2008.* Washington, DC: Department of Commerce, 2008. http://www.census.gov/compendia/statab/2008/2008edition.html.

———. *Statistical Abstract of the United States: 2011.* Washington, DC: Department of Commerce, 2011. http://www.census.gov/prod/2011pubs/11statab/infocomm.pdf.

U.S. News & World Report. "On Horizon: Home Computers with a Gift of Gab (Interview with Howard Anderson)." September 10, 1984, 58–59.

Ubinas, Luis A., and Thomas T. Yang. "Classified Ads: How Newspapers Can Fight Back." *McKinsey Quarterly*, January 2006. http://www.mckinseyquarterly.com/ Classified_ads__How_newspapers_can_fight_back_1726.

Udell, Jon G. *The Economics of the American Newspaper.* New York: Hastings House, 1978.

Underwood, Doug. "Reinventing the Media: The Newspapers' Identity Crisis." *Columbia Journalism Review*, March/April 1992, 24–27.

Van Riper, Tom. "Ink-Stained Wretches." *Forbes.com*, March 13, 2006. http://www. forbes.com/2006/03/13/mcclatchy-newspapers-media-cx_tvr_0313knightridder. html.

Veronis, Christine R. "The Bells Are Ringing: Newspapers and Telephone Companies Remain Adversaries over Information Services, and This Year May Be Critical." *Presstime*, April 1989, 14–16.

Viewtron Magazine. "The 'Electronic Mall' Arrives; All Under One Roof (Yours)." no. 1, 1983, 6.

———. "Merchants by Category." no. 1, 1983, 7.

———. "Viewtron's Roots Traced; From Telephone to TV to Videotex." no. 1, 1983, 22.

Walker, Tom. "Another Role for Turner." *Atlanta Journal-Constitution*, February 1, 1990, D2.

Wall Street Equity Research. "Analyst Study on Gannett and New York Times—The End of an Era and the Start of a New Generation of E-Newspapers." September 2, 2010, press release. http://www.marketwire.com/press-release/Analyst-Study-on-Gannett-New-York-Times-The-End-Era-Start-New-Generation-E-Newspapers-NYSE-GCI-1313334.htm.

Washington, Frank S. "Paper Battles Online Services." *Automotive News*, September 21, 1998. http://www.highbeam.com/doc/1G1-50325578.html.

Watts, Douglas R. "The '80s: Telecommunications." *Presstime*, January 1980, 40–41.

Weingarten, Fred W. "Testimony before Congress: New Information Technology and Copyrights." *Computers & Society* 13, no. 3 (1983): 4–8.

Weintrob, Ed. "An Eerily Prescient 1994 Vision of the 2010 iPad. When Newspaper People Calmly Planned for a Future Their Leaders Didn't Have the Courage to Build." *Coney Media: Envisioning a Future*, March 8, 2010. http://coneymedia. wordpress.com/2010/03/08/an-eerily-prescient-1994- vision-of-the-2010-ipad-when-newspaper-people-calmly-planned-for-a- future-their-leaders-didnt-have-the-courage-to-build/.

Weisenthal, Joseph. "After Layoffs at McClatchy, a Focus on Pruitt's Pay; What Are McClatchy's Digital Goals?" *paidContent.org*, June 17, 2008, http://paidcontent. org/article/419-after-layoffs-at-mcclatchy-focus-on- pruitts-pay/.

Weiss, Walter. "Effects of the Mass Media of Communication." In *The Handbook of Social Psychology*, vol. 5, edited by Gardner Lindzey and Elliot Aronson, 77–195. Reading, MA: Addison-Wesley, 1969.

Wellborn, Stanley. "A World of Communications Wonders." *U.S. News & World Report*, April 9, 1984, 59–62.

Wendland, Mike. "Convergence: Repurposing Journalism." *Poynter Online*, February 26, 2001, http://www.poynter.org/content/content_view.asp?id=14558.

Wetmore, Pete. "Classified Market Heats Up with New AOL Player: NAA Seeks a Standard to Build on Web Strengths; More Buy into Classified Ventures." *NewsInc.*, June 8, 1998. http://www.allbusiness.com/marketing/advertising-print-advertising/712621-1.html.

———. "Classified Ventures: What NCN Taught a Newcomer; On-Line Enterprise Is Owned by Three Media Companies Whose Fingers Are at Arm's Length." *NewsInc.*, April 27, 1997. http://www.allbusiness.com/information/information-services-news-syndicates/657663-1.html.

Whitney, Thomas. "All Signs Point to Home-Computer Revolution." *Presstime*, July 1980, 47–48.

Wicklein, John. "The Scary Potentials for the Overcontrol of Information." *Bulletin of the American Society of Newspaper Editors*, December/January 1982, 13–15.

Wilson, Jean Gaddy, and Iris Igawa. "On Our Minds: Worry, Worry, Worry, or Innovate?" *Presstime*, April 1994, 26–28.

Winseck, Dwayne. "Back to the Future: Telecommunications, Online Information Services and Convergence from 1840 to 1910." *Media History* 5, no. 2 (1999): 137–157.

Wired. "The Web Is Dead." September 2010, 118–122.

Wireless News. "RTG Ventures Enters Mobile App Market." August 25, 2011. http://www.highbeam.com/doc/1P1-196061202.html.

Wirtz, Bernd W. "Convergence Processes, Value Constellations and Integration Strategies in the Multimedia Business." *International Journal on Media Management* 1, no. 1 (Autumn/Winter 1999): 14–22.

Wolf, Michael J. "Media Mergers: The Wave Rolls On." *McKinsey Quarterly*, June 2002. http://www.mckinseyquarterly.com/Media_mergers_The_wave_rolls_on_1173.

Wolfe, Gary. "The (Second Phase of the) Revolution Has Begun." *Wired*, October 1994. http://www.wired.com/wired/archive/2.10/mosaic_pr.html.

Wright, Donald F. "Electronic Publishing: How to Use It and Why." *Presstime*, February 1982, 25.

Wright, Donald K. "The Magic Communication Machine: Examining the Internet's Impact on Public Relations, Journalism, and the Public." Monograph published by the University of Florida Institute for Public Relations, 2001: 1– 67.

Wronski, Richard. "Media Companies Seeking New Turf." *Chicago Tribune*, April 29, 1996, 8.

Wyman, Bob. "The Last Link: Employment Listings Are Another Forfeited Franchise." *CMC Magazine*, June 1996. http://www.december.com/cmc/mag/1996/jun/last.html.

Yahoo Finance. "NASDAQ Composite Interactive Chart." 2010. http://finance.yahoo. com/echarts?s=%5EIXIC+Interactive#chart6:symbol=^i xic;range=19950103,20 100603;indicator=volume;charttype=line;crosshair=o n;ohlcvalues=0;logscale=off.

Yamamoto, Mike. "Legacy: A Brave New World Wide Web." *CNET News.com*, April 14, 2003. http://www.news.com/2009-1032-995680.html?tag=toc.

Yelvington, Steve. "Nostalgia for the New Century Network." *Yelvington.com*, January 16, 2006. http://www.yelvington.com/20060116/nostalgia_for_the_new_ century_network.

Zhang, Mabel. "Online Newspapers Readership Increases." *iMedia Connection*, November 16, 2005. http://www.imediaconnection.com/news/7292.asp.

Zuckerman, Laurence. "Newspapers Balk at Scooping Themselves on Their Own Web Sites." *New York Times*, January 6, 1997. http://www.nytimes.com/1997/01/06/ business/newspapers-balk-at-scooping-themselves-on-their-own-web-sites. html?pagewanted=all.

Index

A

ABC television, 151, 175
About.com, 197–198
Advance Publications Inc., 142
Advertising Age, 192–193
advertising revenue
    cable television and, 76–83
    classified advertising losses,
        163–166
    Internet and evolution of, 21
    mobile applications and, 207–210
    NCN platform for, 142–150
    newspaper-cable partnerships and,
        69–76
    newspapers' loss of, 1–2, 181, 209–
        210, 219–220
    online advertising, 167–170,
        194–203
    peak in 2000 for, 9
    proprietary online systems and, 120–
        121, 125–126

short-term profit *vs.* long-term plan-
        ning and, 72–74, 220–222
    statistics for, 7
    telecommunications competition for,
        92–95
    video programming and, 75–76
    videotext's impact on, 40–41
    web editions of newspapers and,
        194–203
    Yellow Pages, 19–20, 96, 102–103
Affiliated Publications, 49
Albarran, Alan B., 208–209
Alber, Antone F., 38, 49
Alhassan, Abubakar D., 175, 179
Allen, Matthew, 223
American Editor, The, 188
American Journalism Review, 113, 124,
        199, 209
American Newspaper Publishers
        Association (ANPA), 75–78, 106
American Society of Newspaper Editors
        (ASNE), 152

America Online (AOL)
  advertising revenue, 9
  business profile for, 122
  development of, 20, 106
  Internet and, 134–135, 208
  newspaper partnership with, 112–113,
    121, 123–126, 218–219
  portal system and, 161
  Time Warner merger with, 21, 173,
    175, 179, 198, 204
Ameritech, 105
analog technology, abandonment of, 33
Andreessen, Marc, 137
anticipated value of content, as industry
    paradigm, 153–154
Antiope, 27
Apple Computer, 62
Arizona Republic, 150
Ashe, Reid, 54, 56–57
Associated Press
  online content from, 157
  online newspapers and, 44–45
  personal computer research, 63
Atlanta Journal-Constitution
  audiotex services and, 99
  online projects at, 45, 128–129, 150
  profitability at, 199
  telecommunication partnerships
    and, 105
AT&T
  antitrust agreement for, 87–88
  breakup of, 95–101
  competition for newspapers from,
    85–86
  concept trial by, 85
  consent decree signed by, 96–97
  regulatory environment for, 86–89
  rhetoric war with newspapers, 89–95,
    219–220
  videotext projects and, 48–50
audiotex services, 98–99, 141

Aumente, Jerome, 28
Austin American-Statesman, 128

B

Bagdikian, Ben H., 27, 33–34
Bank of America, 102
barriers to entry
  online embargoes and, 156–157
  portal systems and, 163
Batten, James K., 50
Becker, Lee B., 70
Bell Laboratories, 88–89
Bell operating companies
  AT&T breakup and, 95–101
  resurgence of, 101–108
BellSouth, 104–105, 116
Belo Corp., 43–44, 53, 185
Beniger, James R., 32–37
Berners-Lee, Tim, 137
Besen, Stanley M., 65
Better Homes & Gardens, electronic
    content from, 36
Beverly Hillbillies, The (television program)
Blethen, Frank, 101
Boczkowski, Pablo J.
  on Internet growth, 131
  on newspaper business models, 42–43,
    118
  newspaper industry research by, 6
  on newspaper viability, 167, 169–170
  online newspapers discussed by,
    141, 152
  personal computers and, 62–63
  on videotext, 27, 31, 47–48, 57–59, 218
Bogart, Leo, 106
Bontis, Nick, 178–179
Boston Globe
  NCN consortium and, 144
  online edition of, 150–152
  Viewtron system and, 49

Bovee, Warren G., 13
British Post Office, 27
broadcasting industry
    cable television competition for,
        64–65
    mergers in, 138–139
Brooke, Collin G., 3
Brown, Charles, 88–91, 96
Bryant, Jennings, 214
Buffett, Warren, 13–15
bulletin board systems (BBS), 46–47,
        111, 115, 164–166
Burns, J. Christopher, 38
business models in newspaper industry
    convergence and, 193–203
    mainstreaming of Internet and, 137–
        139, 141–142
    market conditions and, 37–43
    mergers and, 175–178
    NCN platform standards and,
        142–150
    new models needed for, 221–222
    portal systems and, 161–163
    proprietary online services and, 117–
        118
Business Week, 147–18

C

Cable News Network (CNN)
    development of, 10, 74
    Time Warner purchase of, 175–176
cable television industry
    advertising revenue for, 9
    computer technology and, 62–63
    economic value of, 79–80
    evolution of market for, 74–76,
        86–87
    exit of newspapers from, 76–83
    history of, 63–65
    interactivity and, 65–66

mergers in, 138–139
newspaper company partnerships,
        66–76, 215–217
newspaper industry strategy for, 5,
        8–9, 19, 61–84
short-term expectations in, 72–74,
        220–222
statistics about, 64
videotext development and, 26
cabletext, 71–76, 104
Canada, videotext in, 27
Capital Cities/ABC, 116
capital investment, cable television's
        dependence on, 68–76
Capital-Journal (Topeka, Kansas), 185
CAPTAIN videotext system, 27
CareerPath.com, 144–145
CBS, Viacom purchase of, 175
Centel telephone company, 53
Chan-Olmsted, Sylvia M., 138–139
Chemical Bank, 102
Chicago Sun-Times, 105
Chicago Tribune, 144, 177, 185
Christian Science Monitor, 159
Chyi, Hsiang I., 153, 155
circulation revenue
    erosion of, 9–10, 42–43
    newspaper viability and, 168–170
    statistics and trends in, 7
Citigroup, 210
classified advertising
    Internet threat to, 163–166,
        220–221
    online revenue from, 195–203
    videotext's impact on, 40–41
Classified Ventures, 165–166
CLTV, 185
CNET News, 137, 145–146, 150
Cohen, Bernard, 256n.123
Cohen-Avigdor, Nava, 198
Columbia Journalism Review, 178–179

commercial information services
  Bell operating companies and, 101–104
  newspaper partnerships with cable television and, 69–76
  online newspapers and legacy of, 34–37, 141
  telecommunications competition in, 92–95
communication technology
  history of, 2–3
  rhetoric of, 32–34
Compaine, Benjamin M., 11–12, 30, 41, 219–220
CompuServe
  convergence and, 225
  decline of, 175
  development of, 35–37, 106
  Internet and, 134–135
  newspaper partnerships with, 112–113, 121, 123
  profile of, 122
  proprietary online system, 44–46
  videotext systems and, 53
Computer Industry Almanac, 62
computer technology. *See also* personal computers
  cable television and, 62–63
  electronic publishing and, 29–31, 207–210
  historical impact of, 4
  Neuharth's predictions about, 8
  newspaper industry reliance on, 29–31
  predicted impact on newspaper industry of, 10–11
  print-based publishing and, 39–43
  social impact of, 27
  videotext development and, 26
  Viewtron failure and shortcomings of, 57

consumer market
  evolution of cable television and, 64–65, 74–76
  Gateway system and, 50–52, 217–219
  information databases and, 38–43
  newspaper websites and, 194–205
  online newspapers and, 44–46
  portal systems and, 161–163
  Viewtron system and, 49–50, 217–219
content management
  anticipated value paradigm, 153–154
  central distribution model for, 135
  compensation issues, 120–121
  convergence projects and, 185–186
  interactivity in, 135–139, 158–160
  NCN experiment and, 148–150
  newspaper-cable partnerships and, 71–74
  online competition for, 160–166
  online embargoes, 156–157
  separation of technology from, 71–74
  shovelware paradigm and, 153–156
convergence
  current perspectives on, 225–226
  electronic publishing and, 31
  limitations of, 190–193
  Media General experiment and, 185–190
  newspaper industry strategy of, 181–193, 223
  newspaper-television cooperation and, 182–185
  videotext and, 24–25
corporate convergence, defined, 18
cost controls, videotext development and, 42–43
Covidea home banking service, 102
Cox Newspapers Inc. (Cox Enterprises)
  audiotex services and, 99
  cable television and, 80

Cox Interactive Media formation, 150, 191
divestment of online operations, 178
NCN consortium and, 142, 148
operational convergence plans at, 191
Prodigy project and, 127–129
profitability of, 199
proprietary online systems and, 120
telecommunication partnerships and, 105–106
video programming and, 75–76
Craigslist, 196
Crandall, Robert W., 65
Crichton, Michael, 10–11
Criner, Kathleen, 98
cross-media partnerships, 183–186, 190–193
Cuadra, Carlos A., 29
cultural artifacts in newspaper industry
analysis of, 153–160
information control paradigm and, 214–215
innovation suppressed by, 215–217
interactivity and, 224–225
print pre-eminence and, 213–214

D

daily newspapers, decline in numbers of, 9–10
Daily Oklahoman, 185
Dallas Morning News, 43–44, 53, 185
Damsgaard, Jan, 161
Data Times, 36
DeBoer, Lee, 145
Delphi Internet Services Corp., 121–122
Denver Post, 179
device convergence, defined, 18
Dialog Information Services, 35, 36, 103, 145–146
digerati, evolution of, 134–135, 252n.4

Digital City, 124
digital technology
mobile applications and, 207–210
newspaper industry's embrace of, 33
print pre-eminence and, 213–214
directory publishing, classified advertising and, 89–96
Disney Corporation, ABC acquisition, 175
Dispatch (Columbus, Ohio), online newspaper project, 44–45
diversity principle, 94–96, 100, 102–104
Dominick, Joseph R., 18, 182, 183, 226
Donahue, Thomas R., 53
Dow Jones & Co., 36–37, 80, 105, 197–198
Dow Jones News/Retrieval system, 36–37
Dowling, Michael, 181–182
Downes, Edward J., 16
Dunwoody, Sharon, 70
duPlessis, Renee, 184

E

Easterley, David, 99, 106–107
eBay.com, 223
Economist magazine, 210
editorial issues, online technology and, 40–43
Editor & Publisher (trade journal), 31, 209
ElectionLine project, 151
electronic databases, 34–37
electronic library services, 36–37
electronic publishing. *See also* proprietary online services
convergence and, 225–226
early ventures in, 18–19
Internet and abandonment of, 141
newspaper industry complacency regarding, 97–99
newspaper involvement in, 207
videotext and, 28–31, 114–115

electronic tablets
early prototypes for, 114, 116, 141
emergence of, 207–210
eMarketer.com, 208
Emmis Broadcasting, 185
employment levels in newspaper industry
declines in, 10
statistics on, 7
Europe, online technology in, 112
Excite, 161
Eyal, Chaim H., 70

F

Facebook, 209
facsimile systems, 104–105, 141
Fallows, James, 181
family ownership, mergers as threat to,
176–178
fantasy baseball, 105
Federal Communications Commission
(FCC), 65, 73, 86, 88, 91–95
Fedida, Sam, 27
Field Enterprises, 53
Fink, Conrad C., 91
Finnegan, John R., 82
First Amendment
diversity principle and, 94–96, 100,
102–104
influence of, 211–212
newspaper-cable partnerships and
concerns over, 72–74
Flichy, Patrice, 27
Florida Times-Union, 79
Fort Worth Star-Telegram, StarText
project and, 46–47, 52, 115
France
online technology in, 112
videotext in, 27
franchise systems
newspaper-cable partnerships and, 73–74

proprietary online systems and,
120–121
free speech concerns, influence of, 211
Fuller, Keith, 44

G

Gannett Inc.
cable television and, 80–81
computer technology and, 8
Internet businesses purchased
by, 198
NCN consortium and, 142,
146, 148
reorganization of, 207, 209, 222
videotext projects and, 114, 116
Gartner Inc, 192
Gates, Bill, 160
Gateway system
development of, 18–19, 24, 50–52, 79
legacy of, 53–58, 100–101, 217–219
General Electric
GEnie system, 121
NBC acquired by, 7, 175
Gentry, James, 184
Gershon, Richard A., 175, 179
Ghosh, Deb, 49–50
Gilbert, Clark, 194–197, 205
Gillen, Albert J., 23
Gitelman, Lisa, 4, 6
Globe-Gazette (Mason City, Iowa), 69
Gomery, Douglas, 83
Google, 93, 195–196, 198,
204–205, 223
Gordon, Rich, 190
graphical interface, mainstreaming of the
Internet and, 136
Green, Nicola, 16
Greene, Harold C., 95–96, 100–102
Griffin Communication, 185
Gurnsey, John, 29

H

Haddon, Leslie, 16
Harte-Hanks, 80
Hartford Courant, 120
Hartley, John, 17–18
Hearst Newspapers, 116, 142, 148
Hecht, Jeff, 45
Henke, Lucy L., 53
Hinton, Sam M., 161
Hodge, Bob, 3
Holt, Jennifer, 6
home banking service, development
    of, 102
Home Box Office (HBO), 74, 145
Honeywell, 53
Houston Chronicle, 98–99, 152
H&R Block, 35
human-computer interaction (HCI),
    interactive media and, 16–17
Hype Cycle model, 192–193
hypertext, 136–137, 223

I

IBM, 62, 116, 126–129, 152
Igawa, Iris, 116–117
InfiNet project, 168
Infobank database service, 36–37
information databases, consumer market
    potential of, 38–43
information marketplace, model of,
    11–12
Information Society rhetoric
    AT&T-newspaper competition and,
        89–95
    cable television and, 65–66
    Internet and, 135, 229–230
    online project failure and, 217–219
    videotext projects and, 27, 31–36, 40,
        111–113

information technology
    new media and, 222–223
    origin of term, 33
Information Today, 112, 126–127,
        145–146
Institute of Graphic Communication, 29
Institute of the Future, 25
interactive media
    defined, 16–17
    independent newspaper initiatives in,
        150–152
    Internet distribution of, 135–139
    NCN consortium and, 143–150
    newspaper industry's development of,
        19
    Viewtron failure and ignorance of,
        54–58
interactivity
    cable television, 65–66
    defined, 16
    newspapers' acceptance of, 222
    in online newspaper content, 158–160
    paradigm shifts and, 224–225
    StarText project, 46–47
    videotext and, 28
Interchange Network, 121, 122
interim technologies, videotext as, 53–58
International Resource Development, 25
Internet
    advertising revenue on, 194–203,
        208–210
    aftermath of bubble collapse and,
        178–179
    bubble and collapse of, 174–181,
        204
    convergence and impact of,
        193–203
    demise of StarText project and, 47
    disruption of print-based media by,
        133–172, 218–220, 227–230
    hyping of, 23

Internet (*continued*)
  independent newspaper initiatives on,
    150–152
  mainstreaming of, 134–139
  market disruption by, 129–131
  McLuhan's media history and, 2–3
  Mosaic development and, 136–137
  NCN Consortium and, 142–150
  newspaper industry migration to,
    9–15, 20–21, 170–172
  online content embargoes on,
    156–157
  portal systems on, 161–163
  in post-bubble era, 21
  regulatory environment and, 244n.7
Internet Service Provider (ISP) opera-
    tions, newspapers' establishment of,
    168–170

J

Japan, videotext in, 27
Johnson, Robert M., 44, 100–101
Johnson & Johnson company, 41
joint operating arrangements (JOAs),
    245n.22
Jones, Steven, 213
journalism
  cable television and principles of,
    72–74
  convergence resisted by, 187–190
  freedom of the press and influence of,
    211–212
  impact of Knight-Ridder demise on,
    201–203
  mergers as threat to, 176–178
  mission and standards of, 13–15
  operational convergence limitations in,
    190–193
Jung, Jaemin, 192
J. Walter Thompson, 127–128

K

Kaletsky, Anatole, 4
Kapor, Mitchell, 135
Kauffman, Jack, 41
KEYCOM system, 53
KEYFAX system, 53–54
Kilker, Julian A., 159
Kinsley, Michael, 94–95, 112
Kiousis, Spiro, 16
Kist, Joost, 31
Kleiner Perkins Caufield & Byers, 148
Knight-Ridder, Inc.
  AOL partnership with, 124–126
  AT&T partnership with, 93
  commercial databases and, 146
  demise of, 174, 199–203, 205,
    209, 221
  electronic libraries and, 37
  flat-panel electronic tablet project,
    114, 116, 141
  InfiNet project, 168
  information design laboratory, 141
  Internet businesses purchased by, 198
  losses at, 167
  NCN Consortium and, 142,
    145, 148
  newspaper-cable partnerships and,
    72–74
  online newspaper projects and,
    150, 152
  resistance to change in, 57–58
  sale and closing of, 10, 21
  Viewdata Corporation (online subsid-
    iary), 23–24, 27, 54
  Viewtron project, 19, 23–24, 26, 37,
    41, 46–50, 53–58, 100–101, 110–
    111, 114–115, 217–219
Kolodzy, Janet, 186, 188
Koziol, Michael, 15
KSNT-TV, 185

KWTV, 185
Kyrish, Sandy, 3–4, 6, 24–26, 58,
    112–113

L

Laakaniemi, Ray, 45
Lacy, Stephen, 166
Lancaster, F. W., 39
Landmark Communications, 145, 168
Lapham, Chris, 153, 158
Lawson-Borders, Gracie, 183, 185,
    188–189
Leavitt, Harold, 33
Lechner, Christian, 181–182
Leckner, Sara, 157
Ledger-Star (Norfolk, Virginia)
    online projects at, 45
    videotext systems and, 53
Lee, William Chee-Long, 141, 157, 168
LeGates, John C., 92–94
Lehman-Wilzig, Sam, 198
Lerner, Rita G., 29
Lessersohn, James, 104
Lexington Herald-Leader (Kentucky),
    72–74
Li, Xigen, 184
Light, Jennifer S., 5, 104
Lind, Jonas, 192
lobbying activities of newspaper industry,
    20, 91–95
local market control
    media convergence and, 182–186,
        245n.22
    mergers as threat to, 176–178
Los Angeles Times
    audiotex investment by, 98–99
    CareerPath.com and, 144
    electronic information services and,
        36–37
    Gateway system and, 50–51

merger impact on, 177
NCN consortium and, 149
online projects at, 45, 123, 169
Prodigy partnership with, 128–129

M

Machlup, Fritz, 32–33
Manovich, Lev, 15
Mansell, Robin, 161
Marbut, Robert G., 96–97
market conditions
    aftermath of bubble collapse and,
        178–179
    cable television and, 71–76
    Internet and disruption of, 129–131
    Internet bubble and collapse, 174–181
    mergers and alteration of, 175–178
    mobile applications and, 207–210
    newspaper response to Internet and,
        139–152
    videotext systems and, 25–43, 55–58
Martin, Hugh J., 166
Massachusetts Institute of Technology,
    116–117, 216–217
mass market forces, videotext legacy and,
    54–58
Mayer, William G., 15
McCann-Erikson, 116
McClatchy Newspapers, 78, 80–81,
    200–201
McCombs, Maxwell E., 70
McLuhan, Marshall
    Information Society rhetoric and influ-
        ence of, 2–3, 33–34
    new media and, 222–223
McMillan, Sally J., 16
Mead Data Central, 37
media conglomerates
    market conditions and, 175–178
    online newspaper editions and, 150

media convergence
  defined, 17–18
  Media General case study, 21
Media General, 21, 185–190, 226
media history
  current knowledge and context of, 3–4
  industry perspective in, 5–7
  Internet in context of, 2–3
media usage models, newspaper partner-
    ships with cable television and,
    69–76
MediaWeek, 140
Medlars system, 34
Meisel, J. B., 163
mergers
  impact on newspapers of, 173–205
  market alteration and, 175–178
  in media industry, 138–139
Merrill Lynch, 203
Meyer, Phil, 56
Miami Herald, Viewtron project and,
    48–49
Microsoft Corporation, 129, 134,
    160, 198
Microsoft Network, 129, 133
Mill, Jason, 178–179
Mings, Susan M., 165
Minitel system, 112
Minneapolis Star Tribune, online projects
    at, 45
mobile applications, emergence of,
    207–210
*Mobile Communications: An Introduction to
    New Media* (Green and Haddon), 16
Monster.com, 93, 196–197, 223
Moriarity, Gerald, 69
Morris Communications, 185
Mosaic browser
  development of, 20, 133–136, 171
  Internet mainstreaming and,
    136–139

Mowshowitz, Abbe, 34–35, 97
MSN, 160–161
Murdoch, Rupert, 122

N

Naisbitt, John, 8, 33
NASDAQ, Internet stocks on, 178–179
National Cable Television Association
    (NCTA), 65–66, 103
National Library of Medicine, 34
National Science Foundation, 39
NBC, General Electric acquisition
    of, 175
Netscape, 137
Neuharth, Allen H., 8
Neustadt, Richard M., 28–29, 38
New Century Network (NCN) consor-
    tium, 20–21, 134, 141–150, 170–
    171, 177–178, 215–217, 221–222,
    230
new media
  defined, 15–16
  historical context for, 4
  misunderstandings about, 222–223
"News 2000" program, 114
News Center, 186–187, 189–190
Newsday, 105, 125–126
"News in the Future" project, 116–117,
    216–217
News & Observer (Raleigh, North
    Carolina), 129
NewsOK.com, 185
Newspaper Advertising Bureau, 41
newspaper industry
  aftermath of bubble collapse and,
    178–179
  basic characteristics of, 7
  business models in, 37–43
  cable television strategy in, 5, 8–9, 19,
    61–84

commercial services and, 34–37
complacency in, 97–99
convergence in, 173–205, 181–193
cultural artifacts of, 153–160
electronic publishing and, 29–31
exit from cable market by, 76–83
forecasting about, 1–2
historical perspective in, 5–7
impact of Knight-Ridder demise on, 201–203
inaction in, 100–101
independent Internet initiatives in, 150–152
information business *vs.,* 11–12, 152–153
interactive media used by, 17
Internet impact on, 134–139, 194–203
journalism mission *vs.* profit in, 13–15
lobbying by, 91–95
loss of market share in, 7, 9–10, 42–43
mergers and acquisitions in, 173–205
mobile applications and, 207–210
online competition for content and, 160–166
online editions launched by, 139–152
online profits in, 179–181
operational convergence failure in, 190–193
partnerships in cable companies, 66–76
portal systems and, 161–163
predicted demise of, 10–11
profitability *vs.* viability in, 166–170, 198–205
proprietary online service partnerships with, 109–110, 118–132
recent conditions and trends in, 7–15
rhetoric war with AT&T, 89–95
status in early 1990s of, 113–118

survival of, 1–2
systemic change in, 115–117
technological determinism and, 27
telecommunications industry
    competition with, 19–20, 85–108
television partnerships and, 182–186
videotext initiatives in, 43–58
Newspaper Preservation Act (NPA), 245n.22
newsprint prices, videotext development and, 39–40
Newsweek magazine, 151
NewsWorks, 145–147
New York Times
    CareerPath.com and, 144
    facsimile systems and, 104
    Infobank service of, 36–37
    Microsoft partnership with, 160
    NCN and, 143–144, 146, 148
    online projects at, 45, 117–118, 150–152, 155–156, 160–161
    operational convergence at, 191–193
    paywall at, 210
    social networking sites and, 209–210
    video programming and, 75–76
    videotext systems and, 54
    web articles in, 137
    web edition of, 180, 194
New York Times Company
    AOL partnership and, 124–126
    divestment of online operations, 178
    NCN Consortium and, 142, 146, 148
    online losses at, 167, 170
    online profits of, 180
    purchase of Internet companies by, 197–198
Nexis database, 36–37, 145
Nielsen/NetRatings, of newspaper websites, 194
Noam, Eli M., 65–66, 87, 244n.7

"not media" concept, 223
Nynex, 105

O

Ogilvy & Mather advertising agency, 41
Olympic Stain advertising campaign, 41
Omaha World-Herald, 14–15
Online magazine, 35, 136
online media. *See also* proprietary online
    services
  convergence in, 184–186
  embargoes and control of, 156–157
  information marketplace model and, 12
  interactivity in, 158–160, 220
  Internet newspaper editions, 140–152,
    155–157
  news content competition from,
    160–166
  newspaper online projects, 1–2, 5–7,
    21–22, 44–52, 179–181, 211,
    217–219
  print pre-eminence and, 212–214
operational convergence
  defined, 18
  limitations of, 190–193, 204–205,
    216–217
Orbit information database, 35

P

Palm Beach Post, 128
Patten, David A., 82
Pavlik, John, 158–159
pay-per-view systems, cable television
    and, 65–66
paywalls, emergence of, 210
Perren, Alisa, 6
personal computers
  commercial information services
    and, 35–37

Gateway system and, 51–52
  newspaper industry research on,
    62–63
  proprietary online services and,
    110–113
  StarText project and, 46–47
personalization, in online content,
    159–160
Peters, Tom, 4, 11
Pfister, Larry T., 44
Philadelphia Inquirer, 150
Picard, Robert G., 30, 152, 202–203
portal system, online newspapers and,
    161–163, 204–205
portfolio concept in operational conver-
    gence, 191–193
Poynter Institute, 55, 126, 187
premium programming, market evolu-
    tion for, 74–76
Presstime, 80–81, 121, 123, 179
Prestel service, 27, 48
print-based media
  erosion of revenue in, 207–210
  industry commitment to, 212–214
  Internet disruption of, 133–172,
    227–230
  newspaper commitment to, 117–118
  online editions and, 179–181
  technology impact on, 39–43
Prodigy
  decline of, 175
  development of, 9, 20, 108,
    112–113, 225
  Internet and, 134–135, 152, 208
  newspaper partnerships with, 121–
    123, 126–129, 218–219
profitability in newspaper industry
  convergence and, 193–203
  short-term goals *vs.* long-term plan-
    ning, 220–222
  viability *vs.,* 166–170

profitable demise concept, 202–203, 221
proprietary online services
    business models and, 117–118
    development of, 9, 20
    emerging market for, 110–113,
      218–219
    interactivity and, 224–225
    Internet disruption of, 129–131, 134–
      135, 141, 208–210
    multiplicity of options in, 121–129
    newspaper partnerships with, 109–
      110, 118–132
    Viewtron demise and, 49–50

Q

quality control, profitability and, 13–15
QUBE interactive system, 26, 45

R

radio industry
    competition for newspapers from, 7,
      25–26
    information control and, 214–215
    mergers in, 138–139
    newspaper industry investment in, 80
Rafaeli, Sheizaf, 70
Raymond, Darrell R., 54–56
Reader's Digest, 36
Realtor.com, 196
Recorder (Amsterdam, New York), 70–71
regulatory environment
    cable television, 65, 73
    influence of, 211–212
    Internet emergence and, 244n.7
    telecommunications industry in,
      86–89
Reid, Robert H., 129–130
Reuters news service, 157
Rheingold, Howard, 111

rhetoric, in videotext, 32–34
Rice, Ronald E., 15
Ridder, Anthony, 200–201
risk management
    suppression of innovation and,
      215–217
    videotext development and, 26
Rittenhouse, Robert G., 44
RMH Research, 45
Robinson, John, 70
Rocky Mountain News, 105
Rosen, Jay, 202–203, 221
Russell, J. Thomas, 40

S

San Diego Union-Tribune, 181
San Francisco Chronicle, 45
San Francisco Examiner, 45
San Jose Mercury News, 124–126, 144
Schonfeld, Erick, 214
Scott, Ben, 182–183, 190
Sears, 126–129
Seattle Times, 129
Shaw, Rochelle, 161–162
Shawnee News-Star (Oklahoma), 71
shovelware, 153–156
Sidewalk project, 160
Sigel, Efrem, 27–29, 48
Silvie, George, 153, 155
Singer, Jane B., 183–185
Slate online magazine, 198
smartphones, emergence of, 207–210
Smith, Anthony, 7, 32, 87
social contract paradigm in journalism
    information control and, 214–215
    print pre-eminence and, 213–214
Source, The (information database),
    36–37
Southwestern Bell, 105
Sparks, Glenn G., 3

Spokesman-Review (Spokane, Washington), 47
StarText project, 18–19, 24, 46–47, 115
St. Louis Post-Dispatch, 104
   online projects at, 45
St. Petersburg Times, 150
Strategic Inc., 25
Ströber, Rudolf, 4
Sullivan, T. S., 163
"superdesk" concept, 187–190

T

TampaBay Online, 186, 189–190
Tampa Tribune, 128–129, 186–190
Tandy Corporation, 46–47
technological determinism, videotext development and, 27
technology
   cable television's dependence on, 68–76
   impact on newspaper industry of, 8–15
   media distribution through, 17
telecommunications industry
   competition with newspaper industry from, 19–20, 85–108, 219–222
   cooperation with newspapers and, 19–20
   mergers in, 138–139
   newspaper inaction concerning, 99–101
   partnerships with newspapers, 104–108, 168–170
   regulatory environment for, 86–89, 212
   technological changes in, 103–104
telegraph industry
   historical perspective on, 5
   newspapers and, 5
Teletel, 27

teletext
   defined, 28
   limitations of, 24
television industry. *See also* cable television industry
   advertising revenue for, 9, 167–170
   competition for newspapers from, 25–26
   information control and, 214–215
   mergers in, 138–139
   newspaper cooperation with, 182–186
Telidon, 27
Thelen, Gil, 187–189
Thielmann, Bodo, 181–182
Time Inc.
   AT&T partnership with, 102
   cable division of, 67, 71–72
Time magazine, 27
TimesLink project, 128–129
Times Mirror Co.
   advertising revenue at, 196
   cable investments of, 80–81
   classified advertising ventures and, 165, 220
   Gateway project and, 50–58, 100–101, 217–219
   merger with Tribune Co., 166–167, 174, 176–177, 204
   NCN Consortium and, 142
   "New in the Future" project, 116
   online losses at, 167
Time Warner
   America Online acquisition of, 21, 173, 175, 179, 198
   mergers and creation of, 175–176
Toffler, Alvin, 8, 33
Topix.net, 198
Tribune Co.
   cable television and, 76, 80–81
   classified advertising ventures and, 165

convergence strategies and, 185
Internet businesses purchased by, 198
NCN Consortium and, 142, 148
online losses at, 167
proprietary online services and, 119,
123–126
Times Mirror merger and, 166–167,
174, 176–177
videotext projects and, 116
Trintex, 126
Turner, Ted, 10
Turner Broadcasting, 10, 175–176
Twitter, 209
TXCN television, 185

U

Udell, John G., 13, 25–26, 29–30, 42,
211
Underwood, Doug, 113–114
usability studies, videotext systems and,
55–58
USA Today
Microsoft partnership with, 160
NCN consortium and, 146, 148
operational reorganization of, 207,
209, 222
proprietary online systems and, 120–
121, 150
user control, videotext development
and, 40

V

viability of newspaper industry, 166–170,
226–230
Viacom, 175
video display terminals (VDT), development of, 29–30
video programming, newspaper industry
ventures in, 75–76

Videotex Industry Association (VIA),
43–44, 234n.3
videotext
birth of online newspapers and, 23–59
cabletext and, 72–74
classified advertising losses and, 164–166
commercial information services and,
34–37
convergence and, 225
electronic publishing and, 28–31
influence of rhetoric in, 32–34
legacy of, 52–58, 217–219
market forces and, 24–43
newspaper industry initiatives in, 5,
8–9, 43–52, 100–101, 104
online newspapers and legacy of,
44–46, 141
predictions for potential of, 25
risk-averse culture and demise of,
215–217
spelling of, 234n.3
StarText project, 46–47
technological determinism and development of, 27, 221–222
terminology concerning, 27–28
Viewdata Corporation, 23–24, 27, 54
Viewtron Magazine, 48
Viewtron system
development and demise of, 18–19,
37, 41, 47–50, 79, 83, 100–101
legacy of, 53–58, 110, 114–115, 149,
217–219
market expectations for, 24–25, 46–47
Virginian-Pilot (Norfolk, Virginia)
online projects at, 45
videotext systems and, 53
virtual communities, 111
Viswanath, Kasisomayjula, 82
voice information services. *See* audiotex
services
Vu/Text, 36

W

Wall Street Journal, 36–37, 105, 148, 159, 197
Warner Cable, 26
Warner Communications, 75
Washington Post
  CareerPath.com and, 144
  online projects at, 14, 45, 151
  proprietary online services and, 111
Washington Post Co.
  classified advertising ventures and, 165
  information databases and, 38
  NCN platform and, 142, 170
  purchase of Internet companies by, 197–198
Weather Channel, 74
websites for newspapers, 182
Weingarten, Fred W., 32
Weiss, Walter, 70
WFAA-TV, 185
WFLA-TV, 186, 188–189
WGN-TV, 185
Wharton Business School, 221
Whisler, Thomas, 33
White, Peter B., 165
Whole Earth 'Lectronic Link (WELL), 111
"Why the Current Business Model Needs to Change," 199

WIBW radio, 185
Wicklein, John, 73–74
Wilkins, Jeffrey, 45
Wilson, Jane, 98
Wilson, Jean Gaddy, 116–117
Winseck, Dwayne, 5
Wired magazine, 133, 135, 208
wire service content, online newspaper editions use of, 157
Wirtz, Bernd W., 183–184
World Company, 185
World Wide Web
  forecasting about, 208–210
  Mosaic browser and, 136–137
  NCN consortium and impact of, 144–150
  newspapers' embrace of, 131–133, 208–210
Wright, Donald, 36–37
WSB-TV/WSB Radio, 150

Y

Yahoo, 160, 161, 196, 223
Yankee Group, 55
Yellow Pages directories, 19–20, 96, 102–103, 219
Young & Rubicam advertising agency, 41

Digital Formations

General Editor: **Steve Jones**

Digital Formations is the best source for critical, well-written books about digital technologies and modern life. Books in the series break new ground by emphasizing multiple methodological and theoretical approaches to deeply probe the formation and reformation of lived experience as it is refracted through digital interaction. Each volume in **Digital Formations** pushes forward our understanding of the intersections, and corresponding implications, between digital technologies and everyday life. The series examines broad issues in realms such as digital culture, electronic commerce, law, politics and governance, gender, the Internet, race, art, health and medicine, and education. The series emphasizes critical studies in the context of emergent and existing digital technologies.

To order other books in this series please contact our Customer Service Department:

(800) 770-LANG (within the US)
(212) 647-7706 (outside the US)
(212) 647-7707 FAX

To find out more about the series or browse a full list of titles, please visit our website:

WWW.PETERLANG.COM